T0235483

Lecture Notes in Artificial Intelligence 9518

Subseries of Lecture Notes in Computer Science

LNAI Series Editors

Randy Goebel
University of Alberta, Edmonton, Canada
Yuzuru Tanaka
Hokkaido University, Sapporo, Japan
Wolfgang Wahlster
DFKI and Saarland University, Saarbrücken, Germany

LNAI Founding Series Editor

Joerg Siekmann
DFKI and Saarland University, Saarbrücken, Germany

More information about this series at http://www.springer.com/series/1244

Wei Lee Woon · Zeyar Aung
Stuart Madnick (Eds.)

Data Analytics for Renewable Energy Integration

Third ECML PKDD Workshop, DARE 2015
Porto, Portugal, September 11, 2015
Revised Selected Papers

 Springer

Editors
Wei Lee Woon
Electrical Engineering and Computer
  Science
Masdar Institute of Science and Technology
Abu Dhabi
United Arab Emirates

Zeyar Aung
Electrical Engineering and Computer
  Science
Masdar Institute of Science and Technology
Abu Dhabi
United Arab Emirates

Stuart Madnick
MIT Sloan School of Management
Cambridge, MA
USA

ISSN 0302-9743   ISSN 1611-3349   (electronic)
Lecture Notes in Artificial Intelligence
ISBN 978-3-319-27429-4   ISBN 978-3-319-27430-0   (eBook)
DOI 10.1007/978-3-319-27430-0

Library of Congress Control Number: 2015956369

LNCS Sublibrary: SL7 – Artificial Intelligence

This Springer imprint is published by SpringerNature
The registered company is Springer International Publishing AG Switzerland

Preface

This volume contains the papers presented at DARE 2015: The Third International Workshop on Data Analytics for Renewable Energy Integration, which was held in Porto, Portugal, in September 2015 and was hosted by ECML PKDD (the European Conference on Machine Learning and Principles and Practice of Knowledge Discovery in Databases).

Climate change, the depletion of natural resources, and rising energy costs have led to an increasing focus on renewable sources of energy. Much research has been devoted to advancing technologies for extracting energy from these sources. However, a separate concern that is equally important is the efficient and cost-effective storage and distribution of this energy. Even more importantly, a realistic solution to these objectives would also need to be compatible with existing energy infrastructures.

The challenge of renewable energy integration is an inherently multidisciplinary one and depends heavily on robust and scalable computing techniques. In particular, the domains of data analytics, pattern recognition, and machine learning have much to offer in this field. Examples of relevant research topics include the detection of faults, the forecasting of electricity supply and demand, demand response applications, and many others. DARE 2015 provided a forum for researchers working in the various related domains to present and discuss their findings and to share their respective experiences.

We are very grateful to the organizers of ECML PKDD 2015 for hosting DARE 2015, the Program Committee members for their time and assistance, and Masdar Institute of Science and Technology and MIT for their support of this timely and important workshop. Finally, we would also like to thank the authors for their valuable contributions to DARE 2015.

October 2015

Wei Lee Woon
Zeyar Aung
Stuart Madnick

Organization

Program Chairs

Wei Lee Woon	Masdar Institute of Science and Technology, UAE
Zeyar Aung	Masdar Institute of Science and Technology, UAE
Stuart Madnick	Massachusetts Institute of Technology, USA

Program Committee

Osman Ahmed	Siemens Building Technologies, USA
Mohamed Chaouch	United Arab Emirates University, UAE
Tanuja Ganu	IBM Research, India
Oliver Kramer	University of Oldenburg, Germany
Wolfgang Lehner	Techniche Universitäte Dresden, Germany
Depeng Li	University of Hawaii at Manoa, USA
Xiaoli Li	Institute for Infocomm Research, Singapore
Jeremy Lin	PJM Interconnection LLC, USA
David Lowe	Aston University, UK
Bruce McMillin	Missouri University of Science and Technology, USA
Aman Maung Than Oo	Deakin University, Australia
Taha Ouarda	Masdar Institute of Science and Technology, UAE
Pierre Pinson	Technical University of Denmark, Denmark
Roy Welsch	Massachusetts Institute of Technology, USA
Hatem Zeineldin	Masdar Institute of Science and Technology, UAE
Xiangliang Zhang	King Abdullah University of Science and Technology, Saudi Arabia

Contents

Imitative Learning for Online Planning in Microgrids

Samy Aittahar[1]([⊠]), Vincent François-Lavet[1], Stefan Lodeweyckx[2],
Damien Ernst[1], and Raphael Fonteneau[1]

[1] Department of Electrical Engineering and Computer Science,
University of Liège, Liège, Belgium
saittahar@student.ulg.ac.be,
{v.francois,dernst,raphael.fonteneau}@ulg.ac.be
http://www.montefiore.ulg.ac.be
[2] Enervalis, Greenville Campus, Greenville, Belgium
stefan.lodeweyckx@enervalis.com
http://www.enervalis.com

Abstract. This paper aims to design an algorithm dedicated to operational planning for microgrids in the challenging case where the scenarios of production and consumption are not known in advance. Using expert knowledge obtained from solving a family of linear programs, we build a learning set for training a decision-making agent. The empirical performances in terms of Levelized Energy Cost (LEC) of the obtained agent are compared to the expert performances obtained in the case where the scenarios are known in advance. Preliminary results are promising.

Keywords: Machine learning · Planning · Imitative learning · Microgrids

1 Introduction

Nowadays, electricity is distributed among consumers by complex and large electrical networks, supplied by conventional power plants. However, due to the drop in the price of photovoltaic panels (PV) over the last years, a business case has appeared for decentralized energy production. In such a context, microgrids that are small-scale localized station with electricity production and consumption have been developed. The extreme case of this decentralization process consists in being fully off-grid (i.e. being disconnected from conventional electrical networks). Such a case requires to be able to provide electricity when needed within the microgrid. Since PV production varies with daily and seasonal fluctuations, storage is required to balance production and consumption.

In this paper, we focus on the case of fully off-grid microgrids. Due to the cost of batteries, sizing a battery storage capacity so that it can deal with seasonal fluctuations would be too expensive in many parts of the world. To overcome this problem, we assume that the microgrid is provided with another storage

W.L. Woon et al. (Eds.): DARE 2015, LNAI 9518, pp. 1–15, 2015.
DOI: 10.1007/978-3-319-27430-0_1

technology, whose storage capacity is almost unlimited, such as for example as it is the case with hydrogen-based storage. However, this long-term storage capacity is limited by the power it can exchange.

Balancing the operation of both types of storage systems so as to avoid at best power cuts is challenging in the case where production and consumption are not known in advance. The contribution presented in this paper aims to build a decision making agent for planning the operation of both storage systems. To do so, we propose the following methodology. First, we consider a family of scenarios for which production and consumption are known in advance, which allows us to determine the optimal planning for each of them using the methodology proposed in [4] which is based on linear programming. This family of solutions is used as an expert knowledge database, from which optimal decisions can be extracted into a learning set. Such a set is used to train a decision making agent using supervised learning, in particular Extremely Randomized Trees [5]. Our supervised learning strategy provides the agent with some generalization capabilities, which allows the agent to take high performance decisions without knowing the scenarios in advance. It only uses recent observations made within the microgrid.

The outline of this paper is the following. Section 2 provides a formalization of the microgrid. Section 3 describes the related work. Section 4 introduces a linear programming formalization of microgrid planning with fully-known scenarios of production and consumption. Section 5 describes our imitative learning approach. Section 6 reports and discusses empirical simulations. Section 7 concludes this paper.

2   Microgrids

Microgrids are small structures providing energy available within the system to loads. The availability of the power strongly depends on the local weather with its own short-term and long-term fluctuations. We consider the case where microgrids have both generators and storage systems and also loads. This section formally defines such devices. It ends with a definition of Levelized Energy Cost within a fully off-grid microgrid.

2.1   Generators

Generators convert any source of energy into electricity. They are limited by the power they can provide to the system. More formally, let us define G as the set of generators, y^g as the supply power limit of $g \in G$ in Wp and η^g the efficiency, i.e. the percentage of energy available after generation, of $g \in G$ and p_t^g the available power from the source of energy. The following inequation describes the maximal power production:

$$p_t^g \eta^g \leqslant y^g. \tag{1}$$

In our work, we consider photovoltaic panels, for which the power limitation is linearly dependent of the surface size of the PV panel which is expressed in m^2.

Let x^g be the surface size of $g \in G$. The total power production by the surface size is expressed as the following equation, for which the constraint expressed above still holds:

$$\phi_t^g = x^g p_t^g \eta^g. \tag{2}$$

2.2   Storage Systems

Storage systems exchange the energy within the system to meet loads demand and possibly to fill other storage systems. They are limited either by capacity or power exchange. We denote these limits by x^{σ_c} and x^{σ_p} respectively, where $\sigma_c \in \Sigma_C$, $\sigma_c \in \Sigma_P$. We denote $\Sigma = \Sigma_P + \Sigma_C$ as the set of both kind of storage systems and η^σ the efficiency of $\sigma \in \Sigma$. We consider also, $\forall \sigma \in \Sigma$, the variables s_t^σ for the storage content, $a_t^{+,\sigma}$ and $a_t^{-,\sigma}$ for the decision variables corresponding discharge and the recharge amount of the storage system at each time t. Dynamics of storage systems are defined by the following equations:

$$s_{s,0}^\sigma = 0, \forall \sigma \in \Sigma, \tag{3}$$

$$s_t^\sigma = s_{(t-1)}^\sigma + a_{t-1}^{-,\sigma} + a_{(t-1)}^{+,\sigma}, 1 \leqslant t \leqslant T - 1, \forall \sigma \in \Sigma. \tag{4}$$

2.3   Loads

Loads are expressed in kWh for each time step t. Power cuts occur when the demand is not met, and the lack is associated with a penalty cost. Formally, we define the net demand d_t as the difference between the consumption, defined by c_t and the available energy from the generators. The following equation holds:

$$d_t = c_t - \sum_{g \in G} \phi_t^g, \forall 0 \leqslant t \leqslant T - 1. \tag{5}$$

Finally, F_t is defined as the energy not supplied to loads, expressed in kWh, by the following equation:

$$F_t = d_t - \sum_{\sigma \in \Sigma} \eta^\sigma (a_t^{+,\sigma} + a_t^{-,\sigma}), \forall 0 \leqslant t \leqslant T - 1. \tag{6}$$

We now introduce in the model the possibility to have several levels of priority demand, defined by the cost associated to power cuts. Let Ψ be the set of priority demands, and F_t^ψ the number of kWh not supplied for the demand priority group $\psi \in \Psi$. Hence, Eqs. (5) and (6) become:

$$d_t = \sum_{\psi \in \Psi} c_t^\psi - \sum_{g \in G} \phi_t^g, \forall 0 \leqslant t \leqslant T - 1, \tag{7}$$

$$\sum_{\psi \in \Psi} F_t^\psi = d_t - \sum_{\sigma \in \Sigma} \eta^\sigma (a_t^{+,\sigma} + a_t^{-,\sigma}), \forall 0 \leqslant t \leqslant T - 1. \tag{8}$$

2.4 Levelized Energy Cost

Any planning of a microgrid given any scenario of production and consumption leads to a cost based on power cuts and initial investment. It also takes into account economical aspects (e.g. deflation). This cost, called the Levelized Energy Cost (LEC) for a fully off-grid microgrid is formally defined below:

$$LEC = \frac{\sum_{y=1}^{n} \frac{I_y - M_y}{(1+r)^y} + I_0}{\sum_{y=1}^{n} \frac{\epsilon_y}{(1+r)^y}}, \tag{9}$$

where

- n = Considered horizon of the system in years;
- I_y = Investments expenditures in the year y;
- M_y = Operational revenues performed on the microgrid in the year y (take into account the cost of power cuts during the year y);
- ϵ_y = Electricity consumption in year y;
- r = Discount rate which may refer to the interest rate or discounted cash flow.

3 Related Work

Different steps of our contribution have already been considered in various applications. We used linear programming for computing expert strategies. This kind of approach have already been discussed in [7–9] with different microgrid formulations. We used supervised learning techniques with solutions provided by linear programming. Cornélusse et al. [2] have considered this approach in the unit commitment problem. Our prediction model needs an additional step to ensure compliance of the policy learned by supervised learning algorithms with constraints related to the system. This step consists to use quadratic programming to postprocess the solution. Cornélusse et al. [1] have considered a similar approach.

Online planning for microgrids has also been studied with others microgrid configurations. For example, Debjyoti and Ambarnath [3] focuses on the specific case of online planning using automatas, while Kuznetsova et al. [6] focuses on the online planning using a model-based reinforcement learning approach.

4 Optimal Sizing and Planning

4.1 Linear Program

Objective Function. Let $k_t^{\psi}, \forall 0 \leqslant t \leqslant T - 1, \forall \psi \in \Psi$ be the value of loss load of the priority demand $\psi \in \Psi$. The LEC is instantiated in the following way:

$$LEC = \frac{\sum_{t=1}^{T} \frac{-\sum_{\psi \in \Psi} k_t^{\psi} F_t^{\psi}}{(1+r)^{y'}} + I_0}{\sum_{y=1}^{n} \frac{\epsilon_y}{(1+r)^y}}, \tag{10}$$

where $y' = t/(24 \times 365)$.

Constraints. Storage systems actions are limited by their sizes. The following constraints are added:

$$s_t^{\sigma_c} \leqslant x^{\sigma_c}, \forall \sigma_c \in \Sigma_C, 0 \leqslant t \leqslant T-1, \tag{11}$$

$$a_t^{+,\sigma_p} \leqslant x^{\sigma_p}, \forall \sigma_p \in \Sigma_P, 0 \leqslant t \leqslant T-1, \tag{12}$$

$$-a_t^{-,\sigma_p} \leqslant x^{\sigma_p}, \forall \sigma_p \in \Sigma_P, 0 \leqslant t \leqslant T-1. \tag{13}$$

Figure 1 shows the overall linear program.

$$\text{Min.} \quad \frac{\sum_{t=1}^{T} \frac{-\sum_{\psi \in \Psi} k_t^{\psi} F_t^{\psi}}{(1+r)^y} + I_0}{\sum_{y=1}^{n} \frac{\epsilon_y}{(1+r)^y}}, y = t/(365 \times 24) \tag{14a}$$

$$\text{S.t., } \forall t \in \{0 \ldots T-1\} : \tag{14b}$$

$$s_t^{\sigma} = s_{t-1}^{\sigma} + a_{t-1}^{-,\sigma} + a_{t-1}^{+,\sigma}, \forall \sigma \in \Sigma, \tag{14c}$$

$$s_t^{\sigma_c} \leqslant x^{\sigma_c}, \forall \sigma_c \in \Sigma_C, \tag{14d}$$

$$a_t^{+,\sigma_p} \leqslant x^{\sigma_p}, \forall \sigma_p \in \Sigma_P, \tag{14e}$$

$$a_t^{-,\sigma_p} \leqslant x^{\sigma_p}, \forall \sigma_p \in \Sigma_P, \tag{14f}$$

$$\sum_{\psi \in \Psi} F_t^{\psi} \leqslant -d_t - \sum_{\sigma \in \Sigma} \eta^{\sigma}(a_t^{-,\sigma} + a_t^{+,\sigma}), \tag{14g}$$

$$\sum_{\psi \in \Psi} F_t^{\psi} \leqslant 0, \tag{14h}$$

$$-F_t^{\psi} \leqslant c_t^{\psi}. \tag{14i}$$

Fig. 1. Overall linear program for optimization.

4.2 Microgrid Sequence

When planning is performed given sequences of production and consumption of length $T > 0$, a sequence of storage contents and a sequence of actions are generated. Such a group of four sequences is called a *microgrid sequence*. In the following, we abusively denote as an optimal microgrid sequence a set of sequences obtained by solving linear programs. Figure 2 shows an illustration of a sequence of decision, with the two kinds of storage systems. A microgrid sequence is formally defined below:

$$(c_0 \ldots c_{T-1}, \phi_0 \ldots \phi_{0 \ldots T-1}, s_0^{\sigma} \ldots s_{T-1}^{\sigma}, a_0^{\sigma} \ldots a_{T-1}^{\sigma}), \tag{15}$$

where $\forall t \in \{0 \ldots T-1\}, a_t^{\sigma} = a_t^{+,\sigma} + a_t^{-,\sigma}$.

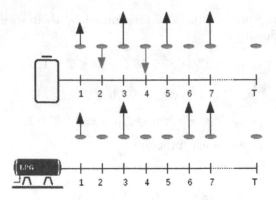

Fig. 2. Sequence scheme (discharging/recharging).

The cost associated to a microgrid sequence is defined below for any microgrid sequence s given any microgrid configuration M, any sequence of production $\phi_{0...T-1}$ and any sequence of consumption $c_{0...T-1}$:

$$LEC_M^{c_0...c_{T-1},\phi_0...\phi_{T-1}}(s) = \frac{\sum_{t=1}^{T} \frac{-\sum_{\psi \in \Psi} k_t^{\psi}((c_t - \sum_{g \in G} \phi_t^g) - \sum_{\sigma \in \Sigma} \eta^{\sigma}(a_t^{+,\sigma} + a_t^{-,\sigma}))}{(1+r)^{y'}} + I_0}{\sum_{y=1}^{n} \frac{\epsilon_y}{(1+r)^y}}. \tag{16}$$

5   Imitative Learning Approach

Optimal microgrid sequences are generated as an expert knowledge database. The decision-making agent is built using a subset of this database. Such an agent is evaluated on a distinct subset.

5.1   Data

Given production and consumption sequences, we can generate microgrid sequences by solving linear programs.

Formally, let $(\phi_t^{(k)}, c_t^{(k)})_{t \in \{0...T-1\}, k \in \{0...K\}}$ be a set of production and consumption scenarios, with $K \in \mathbb{N} \backslash 0$. To this set corresponds a set of microgrid sequences:

$$(\phi_t^{(k)}, c_t^{(k)}, s_t^{(k,\sigma)}, a_t^{(k,\sigma)})_{k \in K, t \in \{0,...,T-1\}}. \tag{17}$$

5.2   From Data to Feature Space

For each time $t \in \{0, \ldots, T-1\}$, production and consumption data are known from 0 to t. Let $\phi_{0...t}^{(k)} = \left\langle \phi_0^{(k)} \ldots \phi_t^{(k)} \right\rangle$ and $c_{0...t}^{(k)} = \left\langle c_0^{(k)} \ldots c_t^{(k)} \right\rangle$ be the sequences of production and consumption from 0 to t. We define a *microgrid vector* from the previous sequences:

$$(\phi_{0\ldots t}^{(k)}, c_{0\ldots t}^{(k)}, s_t^{(k,\sigma)}, a_t^{(k,\sigma)})_{k\in K, t\in\{0\ldots T-1\},\forall\sigma\in\Sigma}. \tag{18}$$

We now introduce, $\forall t \in \{0\ldots T-1\}$, the function $e : \mathbb{R}^t \times \mathbb{R}^t \times \mathbb{R}^{bt} \to \mathbb{R}^{bt}$ where $b = \#\Sigma$. Such a function builds an information vector from sequences of production and consumption. Let $v = e(\phi_{0\ldots t}^{(k)}, c_{0\ldots t}^{(k)})$ be the information vector, v_l the l-th component of v and $L > 0$ the size of the vector. Finally we define, from the definition of microgrid vector, a feature space that will be used with supervised learning techniques as below:

$$(v_1\ldots v_L, s_t^{(k,\sigma)}, a_t^{(k,\sigma)})_{k\in K, t\in\{0\ldots T-1\},\forall\sigma\in\Sigma}. \tag{19}$$

5.3 Constraints Compliancy

An additional step is to ensure that the constraints related to the current information of the system are not violated with the actions performed by the decision making agent. A quadratic program is designed to search for closest feasible actions. This program is defined in Fig. 3. We use constraints from Fig. 1 with an extra one defined below which represents the limit of storage system recharging regarding the overall available energy in the system.

$$\sum_{\sigma\in\Sigma} a_t^{*+,\sigma} \leqslant -d_t' - \sum_{\sigma\in\Sigma} a_t^{*-,\sigma}, \tag{20}$$

where d_t' is defined below to take into account only the possible overproduction by the following equation:

$$d_t' = -max(0, d_t). \tag{21}$$

We are going to illustrate the postprocessing part (see also Fig. 4). We will consider three use cases below, with a battery limited in capacity by 11 kWh and a hydrogen tank with a power exchange limit of 7 kWp.

- Underproduction with both storage systems empty. Initial actions are both discharging but since this is not possible, the postprocessing part cancels the actions (Fig. 4 - top);
- Overproduction with hydrogen tank empty and battery containing 7 kWh. Initial actions are both charging. But the production itself does not entirely meet the consumption. Again, there is a projection where only the battery is discharging (Fig. 4 - middle);
- Underproduction with both storage systems are not empty. The actions are both charging but the energy requested does not meet entirely the consumption. As a consequence, the levels of charging of the battery and of the hydrogen tank are decreased by the projection (Fig. 4 - bottom).

$$\text{Min. } (a_t'^{+,\sigma} - a_t^{*+,\sigma})^2 + (a_t'^{-,\sigma} - a_t^{*-,\sigma})^2 - Ft \tag{22a}$$

$$\text{S.t :} \tag{22b}$$

$$s_t^\sigma + a_t^{*-,\sigma} + a_t^{*+,\sigma} \geqslant 0, \forall \sigma \in \Sigma \tag{22c}$$

$$s_t^{\sigma_c} \leqslant x^{\sigma_c}, \forall \sigma_c \in \Sigma_C , \tag{22d}$$

$$a_t^{*+,\sigma_P} \leqslant x^{\sigma_P}, \forall \sigma \in \Sigma_P , \tag{22e}$$

$$- a_t^{*-,\sigma_P} \leqslant x^{\sigma_P}, \forall \sigma \in \Sigma_P , \tag{22f}$$

$$\sum_{\psi \in \Psi} F_t^\psi \leqslant -d_t - \sum_{\sigma \in \Sigma} \eta^\sigma a_t^{*-,\sigma} + \frac{a_t^{*+,\sigma}}{\eta^\sigma} , \tag{22g}$$

$$\sum_{\psi \in \Psi} F_t^\psi \leqslant 0 , \tag{22h}$$

$$- F_t^\psi \leqslant c_t^\psi , \tag{22i}$$

$$- d_t' \leqslant d_t , \tag{22j}$$

$$- d_t' \leqslant 0 , \tag{22k}$$

$$\sum_{\sigma \in \Sigma} a_t^{*+,\sigma} \leqslant -d_t' - \sum_{\sigma \in \Sigma} a_t^{*-,\sigma} . \tag{22l}$$

Fig. 3. Quadratic program defined for any time step $t \in \{0 \ldots T - 1\}$ (postprocessing part).

5.4 Evaluation

An evaluation criterion consists to compute the difference of cost observed between control by the imitative agent and the optimally control microgrid, for a given context (i.e. profile of production/consumption and microgrid settings). More formally, let's consider s^* the optimal microgrid sequence and s' the microgrid sequence generated by the decision making agent. Then the cost difference is represented by the function below:

$$Err_M^{c_0 \ldots cT-1, \phi_0 \ldots \phi T-1}(s') = LEC_M^{c_0 \ldots cT-1, \phi_0 \ldots \phi T-1}(s') - LEC_M^{c_0 \ldots cT-1, \phi_0 \ldots \phi T-1}(s^*). \tag{23}$$

6 Simulations

6.1 Implementation Details

The programming language Python[1] was used for all the simulations, with the library Gurobi[2] for optimization tools and scikit-learn[3] for machine learning tools.

---

[1] www.python.org.
[2] http://www.gurobi.com/.
[3] www.scikit-learn.org.

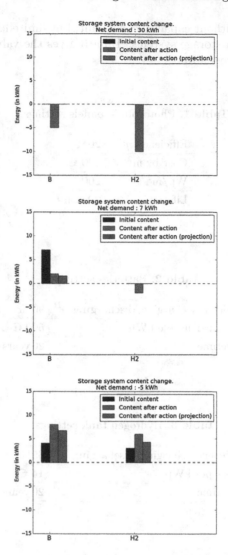

Fig. 4. Bar plots representing the projection from actions to feasible ones when needed.

6.2 Microgrid Components

Devices below are considered for the microgrid configuration.

- Photovoltaic panels accumulate energy from solar irradiance with a ratio of loss due to technology and atmospheric issues. According to Sect. 2, they are defined in terms of m^2 and of Wp per m^2. Table 1 gives the values of the elements describing the PV panels.
- Batteries are considered as short-term storage systems with no constraint on power exchange but with limited capacity. Table 2 gives the values of the elements describing the batteries.

– Hydrogen tanks, without capacity constraint, but limited in power exchange. They are long-term storage systems. Table 3 gives the values of the elements describing the hydrogen tanks.

Table 1. Photovoltaic panels settings.

Efficiency η^{PV}	20 %
Cost by m^2	200 €
Wp/m^2	200
Lifetime	20 years

Table 2. Batteries settings.

Efficiency charging/discharging η^B	90 %
Cost per usable kWh	500 €
Lifetime	20 years

Table 3. Hydrogen tank settings.

Efficiency charging/discharging η^B	65 %
Cost per kWp	14 €
Lifetime	20 years

6.3   Available Data

Consumption Profile. An arbitrarily pattern was designed as a representative model of a common residential daily consumption with two peaks of respectively 1200 and 1750 W. Figure 5 shows the daily graph of such a consumption profile.

Production Profile. The production scenarios are derived from the production data of a photovoltaic panel installation located in Belgium. These data have been processed in a straightforward way so as to have histories of production per m^2 of PV panels installed. These will be used later to define the production scenarios by simply multiplying them by the surface of the PV panels of the microgrid.

Figure 6 shows a typical production scenario for PV panels in Belgium.

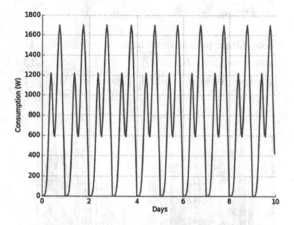

Fig. 5. Residential consumption profile.

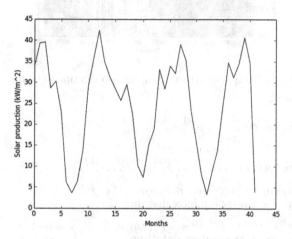

Fig. 6. Monthly production profile of PV panels in Belgium.

6.4   Test Protocol

We split the set of scenarios of production and consumption into two subsets, a learning set for training the agent and a test set to evaluate the performances. The learning set contains the two first years of production and the test set contains the last year of production. We also apply linear transformations as below to artificially create more scenarios for both learning and test sets.

$$\bigcup_{i\in\{0.9,1,1.1\}} \bigcup_{j\in\{0.9,1,1.1\}} \{(ic_t, j\phi_t)\}, t \in \{0\ldots T-1\}. \qquad (24)$$

Table 4 details the configuration of our microgrid.

The following information vectors have been considered, $\forall t \in \{0\ldots T-1\}$:

Table 4. Microgrid configuration.

Photovoltaic panels area (in m^2)	42
Battery capacity (in kWh)	13
Hydrogen network available power (in kWp)	1

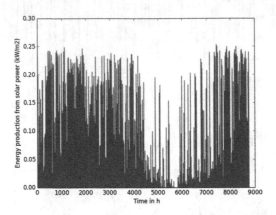

Fig. 7. Sample of typical Belgium production (1 year).

– 12 h of history, i.e. $e(\phi_0^{(k)} \ldots \phi_t^{(k)}, c_0^{(k)} \ldots c_t^{(k)}) =$
$(\phi_{max(t-12,0)}^{(k)} \cdots \phi_t^{(k)}, c_{max(t-12,0)}^{(k)} \cdots c_t^{(k)})$;
– 3 months of history, i.e. $e(\phi_0^{(k)} \ldots \phi_t^{(k)}, c_0^{(k)} \ldots c_t^{(k)}) =$
$(\phi_{max(t-24\times30\times3,0)}^{(k)} \cdots \phi_t^{(k)}, c_{max(t-2160,0)}^{(k)} \cdots c_t^{(k)})$;
– 12 h of history + summer equinox distance, i.e.
$e(\phi_0^{(k)} \ldots \phi_t^{(k)}, c_0^{(k)} \ldots c_t^{(k)}) = (\phi_{max(t-12,0)}^{(k)} \cdots \phi_t^{(k)}, c_{max(t-12,0)}^{(k)} \cdots c_t^{(k)}, |t-t^*|)$,
where t^* is the summer equinox datetime.

A forest of 250 trees have been built with the method of Extremely Randomized Trees proposed in [5]. Our imitative learning agent and our optimal agent are also compared with a so-called greedy agent that behaves in the following way.

– If $d_t \geqslant 0$, i.e. if underproduction occurs, storage systems are discharged in decreasing order of efficiency;
– If $d_t \leqslant 0$, i.e. if overproduction occurs, storage systems are charged in decreasing order of efficiency.

The main idea of this greedy agent is to keep as most as possible energy into the system.

6.5   Results and Discussion

Optimal Sequence of Actions. Figure 8 shows the evolution of the storage systems contents given optimal sequences of actions, for a given scenario. The empirical mean LEC over the test set is 0.32€/kWh. The evolution is plotted over 1 year.

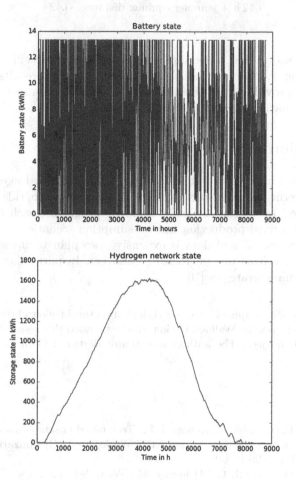

Fig. 8. Storage system state evolution (optimal).

As expected, the battery tries to handle short-term fluctuations. On the other hand, the hydrogen tank content gradually increases during summer before gradually decreasing during winter.

Greedy and Agent-Based Sequences of Actions. Table 5 shows the LEC for all the sequences generated by the greedy algorithm and the controller given several input spaces.

Considering a history of production and consumption of only 12 h is more expensive in terms of LEC, compared to a history of production and consumption

Table 5. Overall mean LECs.

Greedy controller	0.6
12 h	0.44
3 months	0.43
12 h + summer equinox distance	0.42

of 3 months. It shows that a decision making agent is more efficient with long-term information. Additionally, we also report experimental results for which the agent was also provided with the distance (in time) to summer equinox. This additional information improves the performances.

7   Conclusion

In this paper, we have proposed an imitative learning-based agent for operating both long-term and short-term storage systems in microgrids. The learning set was obtained by solving a family of linear programs, each of them being associated with a fixed production and consumption scenario.

As having access to real data is expensive, we plan to investigate how to transfer knowledge from one microgrid to another. In particular, we will focus on transfer learning strategies [10].

Acknowledgments. Raphael Fonteneau is a Postdoctoral Fellow of the F.R.S.-FNRS. The authors also thank the Walloon Region who has funded this research in the context of the BATWAL project. The authors also thank Bertrand Cornelusse for valuable discussions.

References

1. Cornélusse, B., Geurts, P., Wehenkel, L.: Tree based ensemble models regularization by convex optimization. In: 2nd NIPS Workshop on Optimization for Machine Learning, OPT 2009 (2009)
2. Cornélusse, B., Vignal, G., Defourny, B., Wehenkel, L.: Supervised learning of intra-daily recourse strategies for generation management under uncertainties. In: 2009 IEEE Bucharest PowerTech, pp. 1–8. IEEE (2009)
3. Debjyoti, P., Ambarnath, B.: Control of storage devices in a microgrid using hybrid control and machine learning. In: IET MFIIS 2013, Kolkata, India, vol. 5, p. 34 (2013). ISBN: 978-93-82715-97-9
4. Francois-Lavet, V., Gemine, Q., Ernst, D., Fonteneau, R.: Towards the minimization of the levelized energy costs of microgrids using both long-term and short-term storage devices. To be published (2015)
5. Geurts, P., Ernst, D., Wehenkel, L.: Extremely randomized trees. Mach. Learn. **63**(1), 3–42 (2006)
6. Kuznetsova, E., Li, Y.F., Ruiz, C., Zio, E., Ault, G., Bell, K.: Reinforcement learning for microgrid energy management. Energy **59**, 133–146 (2013)

7. Moghaddam, A.A., Alireza, S., Niknam, T., Reza Alizadeh Pahlavani, M.: Multi-objective operation management of a renewable MG (micro-grid) with back-up micro-turbine/fuel cell/battery hybrid power source. Energy **36**(11), 6490–6507 (2011)
8. Morais, H., Kdr, P., Faria, P., Vale, Z.A., Khodr, H.: Optimal scheduling of a renewable micro-grid in an isolated load area using mixed-integer linear programming. Renew. Energy **35**(1), 151–156 (2010)
9. Motevasel, M., Reza Seifi, A., Niknam, T.: Multi-objective energy management of CHP (combined heat and power)-based micro-grid. Energy **51**, 123–136 (2013)
10. Pan, S.J., Yang, Q.: A survey on transfer learning. IEEE Trans. Knowl. Data Eng. **22**(10), 1345–1359 (2010)

A Novel Central Voltage-Control Strategy for Smart LV Distribution Networks

Efrain Bernal Alzate[✉], Qiang Li, and Jian Xie

Institute of Energy Conversion and Storage, University of Ulm, Ulm, Germany
ingebernal@ieee.org, liqiangaps@163.com, jian.xie@uni-ulm.de

Abstract. With the inclusion of Information and Communication Technology (ICT) components into the low-voltage (LV) distribution grid, some measurement data from smart meters are available for the control of the distribution networks with high penetration of photovoltaic (PV). This paper undertakes a central voltage-control strategy for smart LV distribution networks, by using a novel optimal power flow (OPF) methodology in combination with the information collected from smart meters for the power flow calculation. The proposed strategy can simultaneously mitigate the PV reactive power fluctuations, as well as minimize the voltage rise and power losses. The results are very promising, as voltage control is achieved fast and accurately, the reactive power is smoothed in reference to the typical optimization techniques and the local control strategies as validated with a real-time simulator.

Keywords: Smart grid · Renewable energy integration · PV power forecasting · Smart meter · Power system state estimation · LV distribution networks · ICT

1 Introduction

As the penetration of residential and commercial PV increases into electrical distribution systems, an actual trend in the south of Germany, problems such as voltage rise, overloading of network equipment, harmonic current emissions, network resonance, false islanding detection, and dc current injections are becoming more of an issue to be addressed carefully [1]. Concrete solutions for secure and reliable integration of distributed PV generation into distribution systems are a fundamental concern for both academia [4, 6, 7] and industry [8, 19, 32].

A wide range of research projects have focused their efforts particularly on studying how to develop reverse power flow or voltage rise regulation methods to allow this integration. Conventional control mechanisms for distribution voltage regulation are: Voltage control on the feeder by using on-load tap-changing transformers (OLTC) [28, 33, 34], and fixed or switched capacitors to offset the reactive power demand from the load and thus reduce the current flow through the feeder and the related voltage drop [5, 19, 35]. The problem with OLTC transformers or voltage regulators is that, the amount of permissible voltage increase is limited if there is a load near the voltage regulator, a common case in LV distribution networks, so additional voltage regulators along the feeder may be necessary.

© Springer International Publishing Switzerland 2015
W.L. Woon et al. (Eds.): DARE 2015, LNAI 9518, pp. 16–30, 2015.
DOI: 10.1007/978-3-319-27430-0_2

On the other hand, it is also questionable whether the capacitor bank technology is sufficient to answer these challenges, because it may require faster and more flexible control systems than the achievable with capacitor banks [35].

Another potential solution is the use of PV inverters' reactive power as a promising inexpensive concept to resolve the problems caused by PV penetration. Its development and realization attract research efforts in a fairly large number of issues ranging from modeling [3, 8] to implementation [32, 36].

Originally, researchers focused their attention on local or decentralized voltage-control approaches. Nonetheless, in the last few years, optimization techniques to support central control strategies have been proposed, using deterministic optimization methods [13, 24, 25, 27]; non-deterministic optimization methods [20, 23]; and hybrid methods [26, 31].

Central control strategies are demonstrated to be able to resolve voltage violation in LV distribution systems. However, it has repeatedly been shown that these methods may produce unwanted reactive power fluctuations [10, 25, 27, 28]. If the PV penetration is large and widespread, this may also affect subtransmission and transmission systems. This can have important economic impacts and technical implications for distribution substations and transmission lines, such as increasing losses and line loading and so on [9].

At this aim, in this paper a novel central voltage-control strategy is proposed. It is based on optimal reactive power control of smart three–phase solar inverters and the analysis of the data received from the smart meter to solve the OPF. The proposed optimal formulation, which simultaneously minimizes the magnitude of the voltage rise and reduces the power losses, includes a function to smooth the reactive power output of the inverters to improve the power quality in LV distribution network. The remainder of this paper is organized as follows. Section 2 presents the structure of the PV control strategy. The experimental setup and simulation results are illustrated in Sect. 3. Section 4 describes the conclusions drawn from the study carried out in this paper and suggests some guidelines for the future work.

2 PV Control Strategy

The main strategies to control PV systems can be classified as: Local, decentralized and central control strategies [29].

- Local control schemes (also known as droop-based regulation strategies) make autonomous control of the reactive power supply via characteristic curves.
- Decentralized control is based on the control of the reactive power of PV and the interaction with the OLTC transformer in the substation. In this case, some local communication is necessary to enable the interaction between the inverters and the decentralized methodology.
- The central control scheme can be described as a communication based control methodology that allows optimizing the LV distribution grid operation not only locally but also regionally with a common beneficial level for producers and consumers.

2.1 Local Voltage-Control Strategies

The local voltage-control strategies analyzed in this paper correspond to the proposed by German code of practice GC VDE-ARN 4105:

- Power factor characteristic: cosphi(P) method .
- Reactive power/voltage characteristic: Q(U) method.

The cosphi(P) and Q(U) scheme are based on droop characteristic, as shown in Fig. 1. In Germany, the cosphi (P) method proposes a procedure for the calculation of reactive power as a function of the active power generated by the solar systems. With a low active power of PV, the risk of voltage overshoot is low. The reactive power is then adjusted to zero. Once the real power increases to the half nominal power of the solar system, the reactive power increases linearly until the power factor of 0.95 for the case of PV with reactive power level between 3.68 kVA and 13.8 kVA, the typical residential case. For PV with power delivery > 30kVA the suggested power factor is 0.9. Equations (1) express the cosphi(P) curve, as shown in Fig. 1a.

$$\cos \varphi\,(P) = \begin{cases} \cos \varphi 1, \; P < P_1 \\ \cos \varphi 1 + (1 - \cos \varphi 1)\,(P_1 - P/P_1 - P_2)\,, \; P_1 < P \le P_2 \\ -1 + (1 + \cos \varphi 2)\,(P_2 - P/P_2 - P_3)\,, \; P_2 < P \le P_3 \\ \cos \varphi 2, \; P \ge P_3 \end{cases} \tag{1}$$

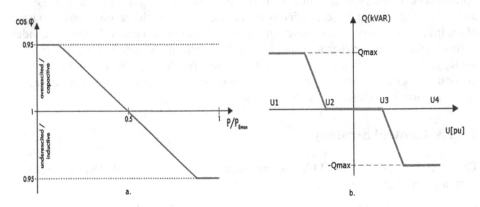

Fig. 1. Characteristic reactive power curves. **a.** cosphi(P) curve. **b.** Q(U) curve.

On the other hand, in the Q(U) method the reactive power of the inverter is regulated as a function of the voltage at the coupling point, as shown in Fig. 1b. It is worth noting that two droop ratios are available when the voltage is higher than the normal range. Besides, achieving better voltage control functions can differentiate the voltage responses of the inverters near LV transformer from the rest along the feeder, so the reactive power contributions from all inverters along the feeder can be more equally distributed as in the case of the cosphi(P) method [10]. As stated in the German GC, the droop curve for the Q(U) method is provided by the network operator.

The algorithm for Q(U) method can be summarized by (2).

$$Q(U) = \begin{cases} Qmax, U < Umin \\ (U - U_1/Umin - U_1).Qmax, Umin \le U < U_1 \\ 0, U_1 \le U \le U_2 \\ -(U - U_2/Umax - U_2).Qmax, U_2 < U \le Umax \\ -Qmax, U > Umax \end{cases} \quad (2)$$

2.2 Central Voltage-Control Strategies

The technical effectiveness of local voltage-control strategies has been very well studied. In particular, study [2] finds that many inverters have the capability of providing reactive power to the grid in order to reduce the voltage rise. Using a similar method, study [3, 6] find that a decentralized voltage-control strategy works just as well, via special located measurement systems, distribution OLTC transformers and controllable PV inverters.

In contrast, central control strategy aims for coordinated control of the complete LV distribution system by using the static and dynamic system information. The target of this strategy is to find the OPF of the LV distribution systems with high penetration of PV. OPF has been the predominant method for such analysis since its introduction by Carpentier (1962) [12].

OPF seeks to optimize a given cost, planning, or reliability objective by controlling the power within an electrical network without violating network power constraints or system and equipment operating limits. Such as conventional power analysis, OPF determines voltage, current, and injected power throughout an electrical power system, that is, the system's state of operation. The general OPF problem is a nonlinear, non-convex, large-scale optimization problem which may contain both continuous and discrete control variables [11].

The most common OPF objective function for the case of PV integration are: Power loss minimization (PLM) [14–16], Voltage rise minimization (VRM) [17–19], PV generation cost minimization (GCM) [20–22], and their combination as multi-objective OPF problems [10, 13, 24, 25, 27].

In the case of VRM, it can be implemented with local control approaches, but the line impedances data is required for the calculation. As when it is implemented in a central control scheme, this approach becomes a similar optimization problem as the PLM strategy [13]. So, this formulation is used as reference to compare the improvements of the proposed central voltage-control strategy, as follows.

2.3 Proposed Central Voltage-Control Strategy

It is well known that central control strategies require the information of the grid topology and the characteristics of the distribution systems, as well as the current status of the buses in terms of voltage, reactive and active power. As the dynamic information is usually only available for a few locations in the system, some distributed state estimation algorithms are required to guarantee the power flow calculation [29, 31]. However, few studies have integrated the actual ICT components of the smart LV distribution networks into the central control strategy [13, 27]. Smart meter is one of these ICT components, which is possible to avoid the necessity of complex state estimation algorithms.

The analysis of the information received from the smart meters allows the power flow calculation of the smart LV distribution system, as presented by the authors in [30] for the implementation of OPF for the central voltage-control methodology. To do so, the proposed control system computes the reactive reference values for the controllable PV inverters every 10 s by minimizing a multi-objective problem using a sequential quadratic programming algorithm developed in Matlab®.

The multi-objective function consists in three optimization objectives, as follows.

Minimization of the power losses:

$$F_1 = \sum_{i=1}^{n-1} \sum_{j=1}^{n-1} R_{ij} S_{ij}^2 / U_{ij}^2 \tag{3}$$

where $i,j = 1,2,\ldots,n$ is the bus number, R_{ij}, S_{ij} and U_{ij} are the resistance of the branch between nodes, the power and voltage obtained from the power flow calculation respectively.

Minimization of the amount of power provided by the PV systems:

$$F_2 = \sum_{i \in \delta} S^2 \tag{4}$$

where δ denotes the set of buses with PV installation [27].

Minimization of the violations to a dead filter-based band to smooth the reactive fluctuations:

$$F_3 = \sum_{i \in \delta} (Q_i - Qf_i)^2 \tag{5}$$

where Q_i is the reactive power for the PV inverter in the bus number, and Qf_i is the output of a Parzen window filter designed to smooth the reactive power fluctuations.

This FIR filter is based on a buffer with the last smoothed reactive power outputs, on which the buffer size corresponds to the size of the average window.

Finally the proposed OPF formulation is as follows:

$$\min \sum_{o=1}^{3} W_o F_o(X)$$

where W_o are the weighting factors for each objective function;

subject to:

$$h = \begin{cases} \left(P_i^2 + Q_i^2\right) \le S_i^2 \\ Umin \le U_i \le Umax \\ |Q_i| \le Q_i max \end{cases}$$

where $Umin$ and $Umax$ denote voltage boundaries and $Q_i max$ corresponds to the maximal reactive power provided by the PV inverter. A basic System schematic of a Smart LV distribution network with the proposed central voltage-control strategy is presented in Fig. 2.

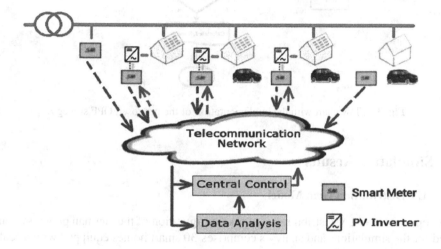

Fig. 2. Smart LV distribution network with central voltage-control strategy.

2.4 Implementation of the Central Voltage-Control Strategy

Figure 3a outlines the architecture of the central strategy developed. For each sample time a power flow analysis is done with the flat start conditions for voltage $U_i = 230 + j0V$, the set active power generation P_i and the reactive power for the PV inverters as $Q_i = 0VAR$. Then, the OPF algorithm takes as reference the output values of the power flow algorithm to find a feasible solution for the defined objective function and the respective restrictions.

Afterwards, the output values of the OPC algorithm are given as reference to the inverters and other load flow analysis is done to check the status of the system and giving a feedback to the OPF algorithms for the smoothing function.

Figure 3b shows the flow chart of the proposed formulation for power flow calculation based on the analysis of the information provided from the smart meters. More detailed description of the power flow calculation with the smart meter information can be found in [30].

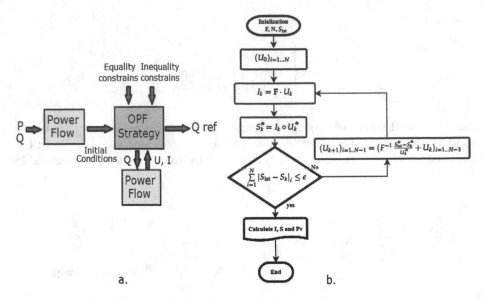

Fig. 3. Flow chart with the implementation of the proposed OPF strategy.

3   Simulation Results

3.1   LV Distribution Grid Model

The representative LV distribution network configuration of the German power system used for the simulations and analyses comprises 30 smart homes equipped with a scalable PV system. All main feeder cables are of type NYY 4×25 mm^2. The households have a three-phase connection with a nominal line-to-neutral voltage of 230 V.

The voltage at the secondary side of the transformer is considered to be 235 V during no load, which can be considered as a typical LV transformer tap to avoid low voltages at the end of the feeder. The solar installations are defined at maximum 5.5 kVA including a 10 % overrating, to support a reactive power compensation until 46 %, even when operating with full power generation. The loads are defined as smart homes instead of passive loads to improve the simulation scenarios. The German smart home load profile used in this paper is very well defined in [32].

3.2   Voltage Rise and Fluctuation

The voltage magnitude along feeders at the end of the simulation for the different reactive power control strategies are illustrated in Fig. 4. Due to the insertion of PV and the fact that there is no control law in the reference case, overvoltage occurs over the limit of 10 % as the worst case scenario, as shown in Fig. 4a.

Despite that all the inverters under cosphi(P) strategy contribute equally to the grid voltage support, the overvoltage limit is reached in the same way as the reference case

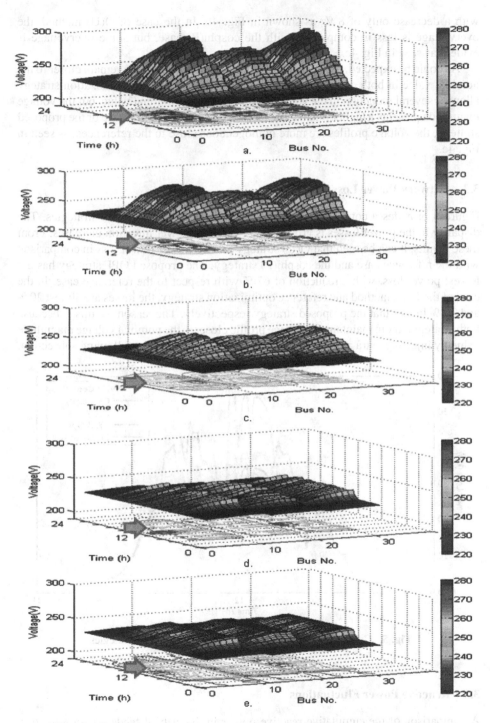

Fig. 4. Comparison of voltage magnitude for the different control strategies.

with a decrease only of 8 %, as shown in Fig. 4b. In the case of Q(U) method, the overvoltage is less in comparison with the cosphi(P) case, but some overvoltage is evident as seen in Fig. 4c.

As could be expected, the bus voltages in the optimization cases remain closer to the nominal value in both cases. It can be observed that the typical optimization strategy allows a several reduction in the voltage level around the nominal value, but the voltage magnitude presents a higher fluctuation, as shown in Fig. 4d. In the case of the proposed strategy, the voltage profiles are more homogeneous close to the references, as seen in Fig. 4e.

3.3   Network Power Loss

Figure 5 provides a comparison of the power losses for the different strategies. The disparity in the power losses is gross as the PV active power injection increases. From these results, it can be seen that Q(U) strategy reaches lower power loss in comparison with the reference case and the cosphi(P) strategy. The proposed OPF strategy has the lowest power loss, with a reduction of 67 % with respect to the reference case. In the case of the Q(U) method and reference optimization strategy, the losses are almost 20 % and 30 % higher that the proposed strategy respectively. The reason for this is because both strategies try to maintain the voltage profile closer to the nominal voltage magnitude value by injecting extra reactive power, as can be observed in the following subsection.

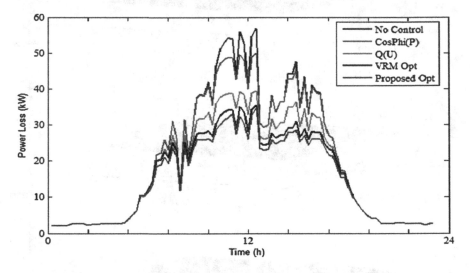

Fig. 5. Comparison of the power losses for the different strategies.

3.4   Reactive Power Fluctuations

A comparison of the cumulative reactive power in the critical feeders is presented in Fig. 6. As expected, in the case of cosphi(P) method, the injected reactive power is

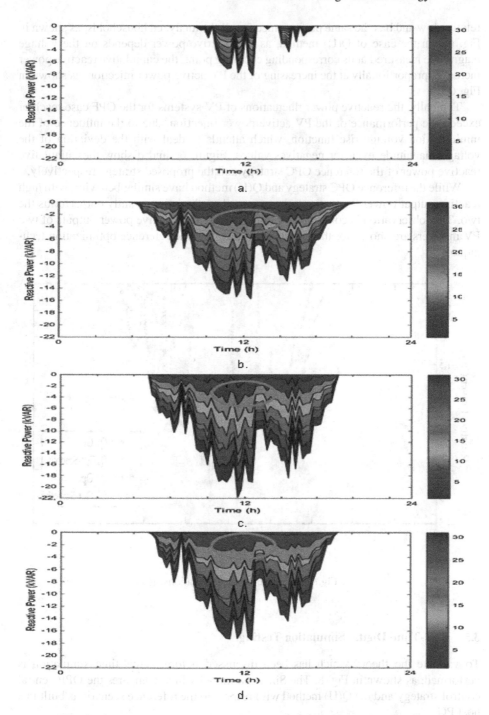

Fig. 6. Comparison of the cumulative reactive power in the critical feeders.

relative low and has the same performance for all the analyzed households, as shown in Fig. 6a. In the case of Q(U) method, as the reactive power depends on the voltage magnitude measured at its corresponding coupling point, the cumulative reactive power increases proportionally at the increasing of the PV active power injection, as shown in Fig. 6b.

Typically, the reactive power fluctuations of PV systems for the OPF case reach or exceed the performance of the PV active power injection, due to the influence of the minimization voltage rise function, which attends to deal with the deviation of the voltage magnitude even for negatives values. Figure 6c and d show the cumulative reactive power of the reference OPC strategy and the proposed strategy respectively.

While the reference OPC strategy and Q(U) method have similar behavior, with high reactive output power fluctuations, the proposed method significantly outperforms the two. This observation is confirmed by Fig. 7, where the reactive power outputs of two PV inverters are shown for the proposed strategy and the reference optimization technique.

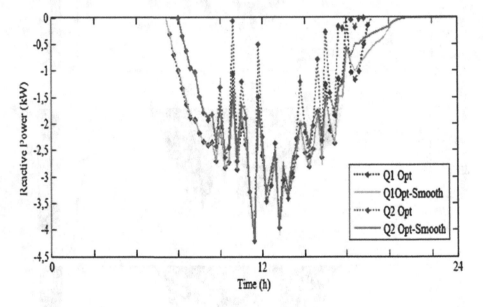

Fig. 7. Reactive power fluctuations.

3.5 Real-Time Digital Simulation Testing

To validate the theory which has been discussed before, a real-time simulation is performed, as shown in Fig. 8. The Simulink model which compares the OPF central control strategy and the Q(U) method with respect to the reference scenario is built in a host PC.

At that point, the code is compiled to the target computer to run the simulation of the smart LV distribution network in real time. Subsequently, the target computer runs the real-time simulation and sends the data, which represents the smart meter's information, to the central controller where the reactive power for the PV inverters is calculated and sends back to the real time simulator.

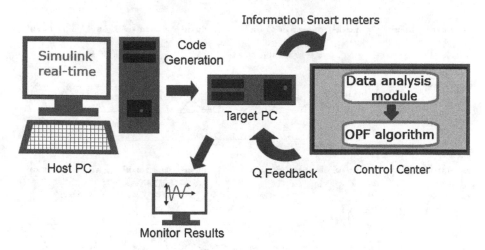

Fig. 8. Real-time simulation scheme based on Simulink Real-Time™.

Figure 9 shows the results obtained from the real-time simulator. Scope 1 shows the PV solar profile simulated for 5 days. Scope 2 and 3 correspond to the RMS voltage response of the household without control nearest and farthest to the transformer respectively.

The results obtained for the smart homes with Q(U) local control strategy are showed in the scopes 4–6, and the smart homes with OPF central control are illustrated in scope 7–9. Scope 4 and 7 compare the reactive power performance of the PV inverter of the household nearest and farthest to the transformer for local and central control respectively.

As can be seen from above real-time simulation results, when the solar power goes higher, the voltage from the household without control exceeds the up limit. The results obtained from the local control improves, but still exceeds the up limit during some peak solar power time period. However, on the other hand, the optimization central control not only regulates the voltage into the limit band for all solar scenarios, but also makes the voltage and reactive power profile smoother in accordance with which has been proposed in this paper.

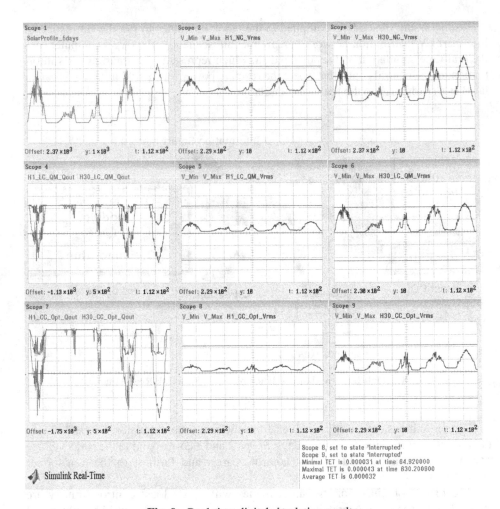

Fig. 9. Real-time digital simulation results.

4   Conclusion

This paper has presented a novel central voltage-control strategy for smart LV distribution networks. The proposed strategy is designed to reduce the reactive power fluctuation of the PV inverters to improve the power quality and reduce the power losses of the system. This is one of the important features of the proposed OPF method, because it improves damping and stability of the reactive power significantly in reference with the typical optimization techniques and the local control strategies.

The results are very promising, as the central voltage-control strategy is capable to mitigate the voltage rise with a better minimization of the power loss. Moreover, a new formulation for the power flow analysis based on the information from the smart meters was included and validated with several simulations included a test with a real-time simulator.

References

1. Mihet-Popa, L., Han, X., Bindner, H., Pihl-Andersen, J., Mehmedalic, J.: Development and modeling of different scenarios for a smart distribution grid. In: IEEE 8th International Symposium on Applied Computational Intelligence and Informatics (SACI), pp. 437–442 (2013)
2. Dimeas, A., Drenkard, S., Hatziargyriou, N., Weidlich, A.: Smart houses in the smart grid: developing an interactive network. Electrification Mag. IEEE 2(1), 81–93 (2014)
3. Casolino, G.M., Di Fazio, A.R., Losi, A., Russo, M.: Smart modeling and tools for distribution system management and operation. In: 2012 IEEE International on Energy Conference and Exhibition (ENERGYCON), pp. 635–640 (2012)
4. Liu, H., Jin, L., Le, D., Chowdhury, A.A.: Impact of high penetration of solar photovoltaic generation on power system small signal stability. In: 2010 International Conference on Power System Technology (POWERCON), pp. 24–28 (2010)
5. Mihet-Popa, L., Han, X., Bindner, H., Pihl-Andersen, J., Mehmedalic, J.: Grid modeling, analysis and simulation of different scenarios for a smart low-voltage distribution grid. In: 2013 4th IEEE/PES on Innovative Smart Grid Technologies Europe (ISGT EUROPE), pp. 6–9 (2013)
6. Tonkoski, R., Lopes, L.A.C.: Coordinated active power curtailment of grid connected PV inverters. IEEE Trans. Sustain. Energy 2(2), 139–147 (2011)
7. Nyborg, S., Røpke, I.: BEnergy impacts of the smart home – conflicting visions. In: Conference Proceedings, Energy Efficiency First: The Foundation of a Low-Carbon Society (ECEEE), pp. 1849–1860 (2011)
8. Austrian Institute of Technology (AT): DG Demonet Smart LV Grid, AT (2011)
9. Katiraei, F., Agüero, J.R.: Solar PV integration challenges. Power Energy Mag. IEEE 9(3), 62–71 (2011)
10. Juanperez, M., Yang, G., Kjær, S.: Voltage regulation in LV grids by coordinated volt-var control strategies. J. Mod. Power Syst. Clean Energy 2, 319–328 (2014)
11. Frank, S., Steponavice, I., Rebennack, S.: Optimal power flow: a bibliographic survey I. Energy Syst. 3, 221–258 (2012)
12. Carpentier, J.: Contribution to the economic dispatch problem. Bull. Soc. Fr. Electr. 8, 431–447 (1962)
13. Kabiri, R., Holmes, D.G., McGrath, B.P.: The influence of PV inverter reactive power injection on grid voltage regulation. In: 2014 IEEE 5th International Symposium on Power Electronics for Distributed Generation Systems (PEDG), pp. 24–27 (2014)
14. Ravindran, V., Aravinthan, V.: Feeder level power loss reduction through reactive power control with presence of distributed generation, pp. 1–25 (2013)
15. Golshan, M.E.H., Arefifar, S.A.: Distributed generation, reactive sources and network-configuration planning for power and energy-loss reduction. IEEE Proc. Gener. Transm. Distrib. 153(2), 127–136 (2006)
16. Yeh, H.-G., Gayme, D.F., Low, S.H.: Adaptive VAR control for distribution circuits With photovoltaic generators. IEEE Trans. Power Syst. 27(3), 1656–1663 (2012)
17. Carvalho, P.M.S., Correia, P.F., Ferreira, L.A.F.: Distributed reactive power generation control for voltage rise mitigation in distribution networks. IEEE Trans. Power Syst. 23(2), 766–772 (2008)
18. Yeh, H.-G., Gayme, D.F., Low, S.H.: Adaptive VAR control for distribution circuits with photovoltaic generators. IEEE Trans. Power Syst. 27(3), 1656–1663 (2012)

19. Cagnano, A., De Tuglie, E., Dicorato, M., Forte, G., Trovato, M.: PV plants for voltage regulation in distribution networks, In: 2012 47th International on Universities Power Engineering Conference (UPEC), pp. 4–7 (2012)
20. Chakraborty, S., Simoes, M.G.: PV-microgrid operational cost minimization by neural forecasting and heuristic optimization. In: Industry Applications Society Annual Meeting, pp. 5–9. IEEE (2008)
21. Budischak, C., Sewell, D., Thomson, H.: Cost-minimized combinations of solar power and electrochemical storage, powering the grid up to 99.9 % of the time. J. Power Sources 232, 15 (2013)
22. Mills, A.D., Wiser, R.H.: Strategies to mitigate declines in the economic value of wind and solar at high penetration in California. Appl. Energy 147, 269–278 (2015)
23. Golestani, S., Tadayon, M.: Distributed generation dispatch optimization by artificial neural network trained by particle swarm optimization algorithm. In: 2011 8th International Conference on the European Energy Market (EEM), pp. 543–548 (2011)
24. Sulc, P., Backhaus, S., Chertkov, M.: Optimal distributed control of reactive power via the alternating direction method of multipliers. IEEE Trans. Energy Convers. 29(4), 968–977 (2014)
25. Kundu, S., Backhaus, S., Hiskens, I.A.: Distributed control of reactive power from photovoltaic inverters. In: 2013 IEEE International Symposium on Circuits and Systems (ISCAS), pp. 249–252 (2013)
26. Boyd, S., Parikh, N., Chu, E.: Distributed optimization and statistical learning via the alternating direction method of multipliers. Found. Trends Mach. Learn. 3(1), 1–122 (2011)
27. Su, X., Masoum, M.A.S., Wolfs, P.J.: Optimal PV inverter reactive power control and real power curtailment to improve performance of unbalanced four-wire LV distribution networks. IEEE Trans. Sustain. Energy 5(3), 967–977 (2014)
28. Liu, X., Aichhorn, A., Liu, L., Hui, L.: Coordinated control of distributed energy storage system with tap changer transformers for voltage rise mitigation under high photovoltaic penetration. IEEE Trans. Smart Grid 3(2), 897–906 (2012)
29. von Appen, J., Braun, M., Stetz, T., Diwold, K., Geibel, D.: Time in the sun: the challenge of high PV penetration in the german electric grid. IEEE Power Energy Mag. 11(2), 55–64 (2013)
30. Alzate, E.B., Mallick, N.H., Xie, J.: A high-resolution smart home power demand model and future impact on load profile in Germany. In: 2014 IEEE International Conference on Power and Energy (PECON), pp. 53–58 (2014)
31. Diwold, K., Yan, W., De Alvaro Garcia, L., Mocnik, L.,Braun, M.: Coordinated voltage-control in distribution systems under uncertainty. In: 2012 47th International on Universities Power Engineering Conference (UPEC), pp. 4–7 (2012)
32. Department of Energy Technology- Aalborg University, Development of a Secure, Economic and Environmentally friendly Modern Power System. DK (2010)
33. Neusel-Lange, N., Oerter, C.: Intelligente Lösungen für Verteilnetze, VDE-Insidel (2012)
34. RWE, Technical University of Dortmund, ABB, and Consentec: Smart Country, Herausforderungen und intelligente Lösungen für die Energiewende auf dem Land., Germany (2011)
35. Turitsyn, K., Sulc, P., Backhaus, S., Chertkov, M.: Options for control of reactive power by distributed photovoltaic generators. Proc. IEEE 99(6), 1063–1073 (2011)
36. Technische Universität München, Georg-Simon-Ohm Hochschule Nürnberg, Siemens AG, and Power Plus Communications AG, Projekt Netz, Germany (2009)

Quantifying Energy Demand in Mountainous Areas

Lefkothea Papada[1(✉)] and Dimitris Kaliampakos[2]

[1] Metsovion Interdisciplinary Research Center, National Technical University of Athens,
9, Heroon Polytechneiou str., 15780 Zographos, Greece
lefkipap@metal.ntua.gr
[2] National Technical University of Athens, 9, Heroon Polytechneiou str.,
15780 Zographos, Greece
dkal@central.ntua.gr

Abstract. Despite their rich energy renewable potential, mountainous areas suffer from energy poverty. A viable solution seems to be the radical turn towards renewable resources. Any tailor-cut energy planning for mountainous areas presupposes the adequate estimation of the energy demand of buildings, which in this case is hindered by the lack of long-term meteorological data, especially in remote, high altitude areas. In this paper four case studies, namely Switzerland, Austria, Greece and north Italy, are examined, applying the method of degree-days. The scarcity of meteorological stations at higher altitudes has been overcome by calculating the lapse rates (decrease of surface temperature with altitude) for each case, which were found to vary from the common "rule" of 6.5°C/km. Based on these findings, the air temperatures of all remote, mountainous spots can be calculated, and, therefore, the estimation of the energy needs of buildings has been provided, with a high level of accuracy.

Keywords: Energy demand · Mountainous · Renewable · Degree-days · Lapse rate · Altitude · Temperature

1 Introduction

The concept of energy crisis lies in the correlation between energy stocks, which tend to decrease, energy consumption requirements which tend to increase and environmental impacts of energy utilization. Fossil fuels reserves are finite, as non-renewable sources, while global energy demand and consumption are constantly increasing. At the same time, the excessive use of fossil fuel resources cause high rates emissions of gaseous pollutants, contributing to climate change. Overall, this trend constitutes a crucial social, economic, political and environmental issue. In this regard, renewable energy resources seem to be the best alternative to the whole energy problem, towards sustainable development.

Mountains are closely related to renewable energy resources. In fact, mountainous terrain along with high-altitude climate, create favorable conditions for the existence of rich renewable energy potential [1, 2]. Specifically, both wind speed and incident solar radiation increase versus altitude, and therefore, mountainous areas are usually rich in wind and solar energy resources. Mountainous areas are richer in water potential as well,

© Springer International Publishing Switzerland 2015
W.L. Woon et al. (Eds.): DARE 2015, LNAI 9518, pp. 31–43, 2015.
DOI: 10.1007/978-3-319-27430-0_3

because of the increasing rainfall and snowfall with altitude. Moreover, many mountain ranges are covered by extensive forests, enriching them with high quantities of biomass.

As a consequence, many opportunities for renewable energy production in mountains are offered. For example, despite the widespread perception of the last decades, that photovoltaics are not suitable for mountainous areas, it is proved that their efficiency increases with altitude, because of the higher incident solar radiation and the prevailing lower temperatures [3]. Water resources, combined with steep slopes of mountainous terrain favor the installation of hydroelectric energy plants. More specifically, regarding hydroelectric power generation, the high drop height of highlands is critical, since it increases the efficiency of the investment [4, 5]. Moreover, forest biomass found in mountains is an important renewable energy resource for thermal energy production.

Overall, it seems that a mountainous area is more possible to have higher renewable energy potential, compared to its nearest lowland [3]. As a result, the exploitation of renewable energy sources (RES) can be the answer to the energy poverty issue, which is known to be a vital problem in mountains, along with a major production activity, boosting local incomes. A representative example is the case of Wildpoldsried, a small mountainous town located in the Bavarian Alps, Germany, at 720 m altitude, which is known as the "energy village". This town began its first RES applications in 1997, mainly through wind turbines and biomass digesters for cogeneration of heat and power. Nowadays, this effort has evolved into a real local industry of solar panels, biogas digesters, windmills, small hydro power plants and a district heating network supplied with biomass. In this way, the town produces five times the energy it consumes, along with a significant annual amount of revenue, being counted up to 4.0 million Euro, while the town's carbon footprint has been reduced by 65 percent.

In any case, the first step for any energy planning within a region is the estimation of its energy needs. The exact determination of the current energy demand is the main prerequisite for the integration of renewable energy into existing energy production systems. However, this is a difficult issue, regarding mountainous areas. The scarcity of meteorological stations at high altitudes (lack of access to long-term meteorological data) has formed a serious obstacle to studying the energy demand variation in mountains. The only adequate network of mountainous meteorological stations is found in high European mountain ranges such as the Alps and in northwestern America. Till now, qualitative estimations about the high energy demand of mountainous areas have been mostly expressed, based on the cold climatic conditions prevailing, not supported by specific quantitative data though. Only a few references, e.g. [1, 6], including quantitative facts have been detected.

In this paper, a methodological tool for quantifying the energy demand of the building sector of four mountainous areas, namely Austria, Switzerland, Greece and north Italy is given.

2 Materials and Methods

Energy demand of buildings has been calculated according to the method of degree-days. The method of degree-days is one of the simplest and most recognized methods for the energy analysis of the building sector [7]. Heating and cooling degree-days (HDDs and CDDs) are quantitative indicators, based on temperature data, designed to reflect the heating and cooling energy needs of a building. Actually, a degree-day symbolizes the quantity and duration at which the external temperature is above or below a defined threshold value, known as base temperature [8–10].

The base temperature is a critical parameter for the degree-days estimation. It is not equal to the interior desired temperature, while it depends on the characteristics of the building examined. However, regarding that technical characteristics vary between different buildings and different areas, a stable value of the base temperature is usually chosen for a specific area, so that its energy demand can be quantified. Since there are no generalized, default values for the base temperature selection, the final choice lies on individual researchers' estimation, taking into consideration the special climatic and temperature conditions of the region examined.

The main advantage of the degree-days methodology is that degree-days are based only on air temperature data. The simplest technique for calculating degree-days uses the difference of the mean daily air temperature from the chosen base temperature [11, 12]. However, since temperature data are usually given on a monthly basis, another method is the most commonly used for calculating degree-days, the Erbs methodology [13], based on monthly temperature data and on the standard deviation of the average daily temperature of the month.

Yet, temperature data, let alone degree-days data, are not usually available in the case of high altitudes, considering the sparse network of mountainous meteorological stations. This difficulty can be overcome by introducing the surface temperature lapse rate value (decrease of surface temperature with altitude). Specifically, lapse rate is produced by quantifying temperature distribution with altitude, thus giving air temperature at a given altitude, in a simple manner. According to the International Standard Atmosphere [14], which regards the Earth's atmosphere as an ideal gas, temperature decreases with altitude at the constant rate of 6.5°C/km, so it can be easily calculated at a given altitude by the following Equation:

$$T = To - 6.5 \times h/1000 \, [°C] \tag{1}$$

where,

T_o: temperature at sea level, defined at 15°C

h: elevation (m)

In this direction, numerous references have been detected, reporting that air temperature of any region can be calculated combining the above atmospheric model with the available temperature data of the nearest meteorological station. The equation then is transformed as follows [15]:

$$T2 = T1 - 6.5 \times (h2 - h1)/1000 \, [°C] \tag{2}$$

where,

T2: temperature of the region examined (°C)

T1: temperature of the nearest region – reference region (°C)

h2: elevation of the region examined (m)

h1: elevation of the nearest region – reference region (m)

This standard lapse rate value (6.5°C/km) has been used in several studies, e.g. [16–19]. However, it has been a subject of dispute between researchers, stating that it varies considerably across different areas. Especially, researchers focusing on mountainous areas by conducting experimental studies with the use of temperature sensor networks, support that the often used values of 6.0--6.5°C/km are not representative of the real conditions, and, even more, that they vary significantly in the case of different mountainous areas, such as the Appalachian mountains [20], the Alps [21], the central Rocky Mountains [22] and elsewhere. In these cases, it is supported that the mean annual lapse rates vary between 3.9°C/km and 5.2°C/km [23]. As a result, defining real lapse rate values in terms of the specific geographical region examined can be very important in energy planning.

After defining a distinct lapse rate value for each one of the cases studied, the mean monthly air temperatures of each remote mountainous station over 600 m are calculated, according to Eq. (2), by applying the air temperature values of the nearest station below 600 m altitude as reference values and the distinct lapse rate value of each region instead of the 6.5°C/km value. In particular, the Eq. (2) is transformed as follows:

$$T2 = T1 - (\text{lapse rate}) \times (h2 - h1) / 1000 \ [°C] \tag{3}$$

Since air temperatures are calculated for any remote mountainous spot, degree-days, and, subsequently, energy needs of the buildings can be then calculated. Heating and cooling energy needs of a building are proportional to the existing climatic conditions, expressed by degree-days, and to the technical characteristics of the building shell, expressed by the heat transfer coefficient. The annual energy demand of buildings for heating is given by Eq. (4) and the corresponding demand for cooling by Eq. (5).

$$Qh = Htot \times HDD \times 24 / 1000 \ [KWh] \tag{4}$$

$$Qc = Htot \times CDD \times 24 / 1000 \ [KWh] \tag{5}$$

where,
Htot: total heat transfer coefficient because both of convection and ventilation (W/°C)
HDD: heating degree-days (°C*days)
CDD: cooling degree-days (°C*days)

Taking into account that the technical features of the building shell vary considerably from one place to another, the heat transfer coefficient included in the energy demand calculation can be kept stable within a geographical area, leaving degree-days as the only factor defining the variation of energy demand from place to place. In this case, the proportion of thermal to cooling energy demand equals to this of HDD to CDD [3]:

$$Qh/Qc = HDD/CDD \tag{6}$$

Furthermore, the variation of degree-days directly reflects the variation of energy needs within a region, by keeping constant the technical characteristics of the buildings. According to [24], the percentage variation of the heating, cooling, as well as the total energy demand of the same building, located at different altitudes, can be calculated:

$$(Qh2 - Qh1) / Qh1 = (HDD2 - HDD1) / HDD1 \ [\%] \tag{7}$$

$$(Qc2 - Qc1)/Qc1 = (CDD2 - CDD1)/CDD1 \ [\%] \tag{8}$$

$$(Qtot2 - Qtot1)/Qtot1 = [(HDD2 + CDD2) - (HDD1 + CDD1)]/(HDD1 + CDD1) \ [\%] \tag{9}$$

To sum up, the main steps of the methodology are:

- Development of a long-term temperature database for all the cases studied, namely Switzerland, Austria, Greece and north Italy, making the analysis more reliable. More specifically, 88 meteorological stations with a 30-year monthly temperature record were used for the case of Switzerland, 63 stations with a 110-year record for the case of Austria, 37 stations with a 107-year record for north Italy and 100 stations with at least a 10-year record for Greece. The meteorological stations list includes a wide range of altitudes. Figure 1 depicts the geographical distribution of the meteorological stations throughout the Alpine region examined.

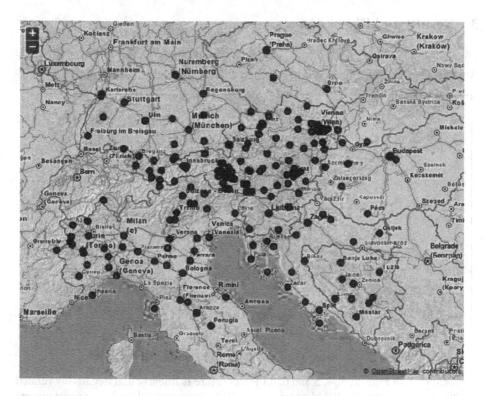

Fig. 1. Geographical distribution of meteorological stations throughout Austria, Switzerland and north Italy (Source: http://www.zamg.ac.at/histalp/dataset/station/osm.php)

- Calculation of the monthly and the annual heating and cooling degree-days for all meteorological stations of the countries examined, according to the Erbs method. The HDDs and CDDs were calculated after selecting the appropriate base temperatures that simulated more realistically each country's climatic conditions. For the case of heating, the base temperatures chosen were 14°C for Switzerland and Austria and 16°C for north

Italy and Greece while for the case of cooling, the base temperatures chosen were 18°C for Switzerland and Austria, 20°C for north Italy and 22°C for Greece.

- Quantification of temperature distribution with altitude for all the cases examined, diagrams and calculation of the corresponding lapse rate values through simple regression models.
- Calculation of air temperatures of all mountainous meteorological stations over 600 m, based on the corresponding lapse rate value and on the air temperature value of the nearest station below 600 m.
- Re-calculation of heating degree-days for the above mountainous stations, based on the new air temperatures calculated. Comparing results with the first ones calculated in step 2.
- Determination of the variation of the energy demand throughout the four regions.

3 Results and Discussion

Temperature distribution with respect to altitude is depicted in Figs. 2, 3, 4, 5 and 6 for Austria, Switzerland, Greece, north Italy and all cases included, respectively.

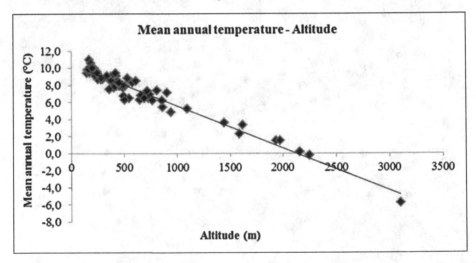

Fig. 2. Temperature distribution with respect to altitude for Austria

As shown in Figs. 2, 3, 4, 5 and 6, air temperature is linearly and negatively related to altitude. More specifically, air temperature decreases versus altitude at a constant rate for all cases. In order to calculate lapse rates, the equations correlating mean annual air temperature with altitude are formed for each country, by performing simple regression analysis. The lapse rate value is the slope of each regression line.

The corresponding equation for Austria is:

$$T = -0.0049 \text{xh} + 10.453 \, [°C] \tag{10}$$

The lapse rate value for Austria is 4.9°C/km.

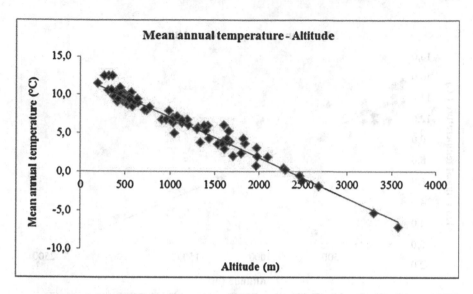

Fig. 3. Temperature distribution with respect to altitude for Switzerland

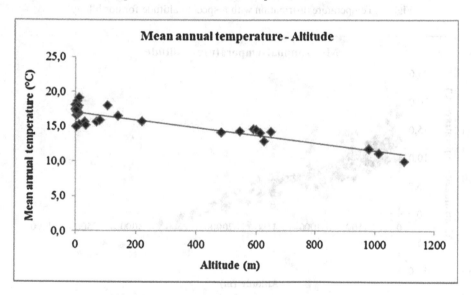

Fig. 4. Temperature distribution with respect to altitude for Greece

The equation for Switzerland is:

$$T = -0.0052 \mathrm{xh} + 12.281 \; [°C] \tag{11}$$

The lapse rate value for Switzerland is 5.2°C/km.

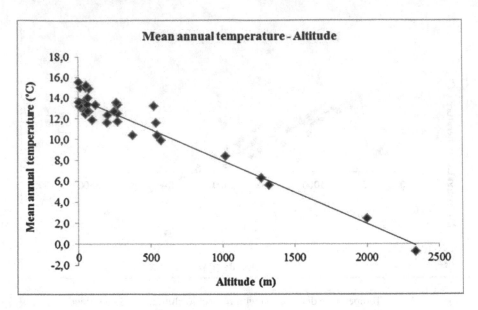

Fig. 5. Temperature distribution with respect to altitude for north Italy

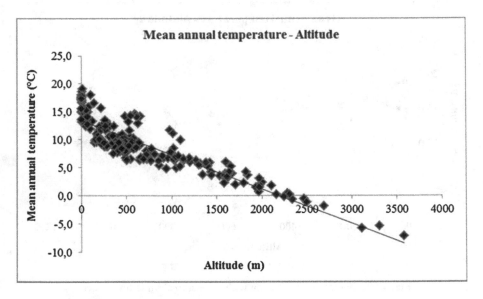

Fig. 6. Overall temperature distribution with respect to altitude for all cases examined

The equation for Greece is:

$$T = -0.0055 \times h + 17.020 \ [^{\circ}C] \tag{12}$$

The lapse rate value for Greece is 5.5°C/km.

The equation for north Italy is:

$$T = -0.0060 \mathrm{x} h + 13.972 \, [°C] \tag{13}$$

The lapse rate value for north Italy is 6.0°C/km.
The equation for the overall distribution examined is:

$$T = -0.0061 \times h + 13.375 \, [°C] \tag{14}$$

The corresponding lapse rate value is 6.1°C/km

Indeed, it is proved that the lapse rate value concerning the sum of cases (6.1°C/km) approaches the common "rule" of 6.5°C/km whereas it varies significantly when focusing on individual mountainous countries or regions, ranging from 4.9°C/km to 6.0°C/km.

The mean monthly air temperatures of each remote mountainous station over 600 m are then calculated, according to Eq. (3). It should be clarified that the Eqs. (10) to (13) can estimate the mean annual air temperature at any location within the countries examined but this annual value is useless regarding the degree-days calculation, and consequently, the energy demand calculation. These calculations require monthly temperature data and the Eq. (3) is the solution to this problem, as allowing the use of monthly values. So, after assessing the monthly air temperatures for the above mountainous stations, monthly and annual heating degree-days are calculated.

Table 1 shows a representative example of the methodology followed, regarding north Italy. More specifically, the monthly air temperatures, as well as the HDDs of the meteorological station "Predazzo" at 748 m were calculated, based on the given air temperatures of the nearest station "Bozen/Bolzano" at 272 m, within 48 km distance. The air temperatures of "Bozen/Bolzano" were used as reference values, and the lapse rate of north Italy used, is equal to 6.0°C/km, according to Eq. (13). The missing data of HDDs in July and August mean that no heating is required during these months. The method gives low deviations between the new calculated HDD values and the "real" HDD values, with a mean monthly value of 12 percent and an annual value of 9 percent. It is noted that the highest value of the HDDs deviation is observed in June, as this month marginally requires heating, and, therefore, the HDDs, even the "real ones" are almost insignificant for this month. As a result, the method provides reliable results for the HDDs in north Italy.

Table 1. Air temperatures and HDDs in Predazzo, Italy, based on the proposed method

	Jan	Feb	Mar	Apr	May	Jun	Jul	Aug	Sep	Oct	Nov	Dec	Annual
T (Predazzo)	-3,8	-1,0	3,8	8,0	12,3	15,8	18,0	17,4	13,7	7,7	1,2	-3,0	7,5
model HDDs	614	475	378	240	122	36	-	-	80	257	443	589	3233
"real" HDDs	531	434	377	257	141	57	-	-	82	207	368	500	2956
HDDs deviation	15%	9%	0%	7%	13%	37%	-	-	2%	24%	20%	18%	9%

Applying this method to all mountainous stations over 600 m in Austria, Switzerland, Greece and north Italy, it is found that the mean deviation between the calculated and the "real" annual HDDs is equal to 7 percent, while this of the monthly HDDs equals to 17 percent, meaning that the method provides very good estimations for the HDDs. Furthermore, it should be noted that only heating degree-days were calculated, since cooling is insignificant at altitudes over 600 m, let alone altitudes over 800 m where cooling tends to be eliminated, as proved by the nearly zero values of the "real" cooling degree-days.

After the HDDs calculation, the variation of energy demand within an area can be assessed. Specifically, given the air temperatures and the HDD values of all meteorological stations within the country, the variation of heating, cooling and total energy needs of the same building located at different altitudes (keeping constant technical characteristics) can be calculated, according to Eqs. 7–9. Moreover, Eq. 6 can give the ratio of heating to cooling needs. An indicative example of the practical use of the methodology suggested is displayed in Table 2, where a random selection of meteorological stations including low and high altitudes was made for Austria, Switzerland, Greece and north Italy.

Table 2. Allocation of energy demand with altitude for a building located in Austria, Switzerland, Greece and north Italy (Cooling is meaningless over 900 m for all cases. The value of 20 CDD at high altitudes has been set as a default value in order to enable calculations. The percentage variation of energy needs at each level has been calculated in respect with the first level (the meteorological station with the lowest altitude))

Country	Meteorol. station	Altit. (m)	Annual HDD (°C*days)	Percent. increase of heat. needs (%)	Annual CDD (°C*days)	Percent. decrease of cool. needs (%)	Percent. variat. of total needs (%)	Ratio of heating / cooling needs (%)	
Austria	Linz-Stadt	263	2357		181			92,9	7,1
	Flattach-Kleindorfl	735	2920	+23,9%	105	-42,2%	+19,2%	96,5	3,5
	Badgastein	1092	3247	+37,8%	20	-89,0%	+28,7%	99,4	0,6
Switzerl.	Lugano	273	1503		410			78,6	21,4
	Chur	556	2101	+39,8%	184	-55,2%	+19,4%	92,0	8,0
	Davos	1594	3591	+138,9%	20	-95,1%	+88,8%	99,4	0,6
North Italy	Bozen/Bolzano	272	2236		260			89,6	10,4
	Brixen/Bressan.	569	2659	+18,9%	134	-48,4%	+11,9%	95,2	4,8
	Predazzo	1020	3233	+44,6%	20	-92,3%	+30,4%	99,4	0,6
Greece	Lamia	143	1192		481			71,2	28,8
	Konitsa	542	1543	+29,4%	249	-48,2%	+7,1%	86,1	13,9
	Karpenisi	980	2074	+74,0%	20	-95,8%	+25,1%	99,0	1,0

As seen in Table 2, heating needs gradually increase versus altitude while cooling needs decrease. Yet, despite the decrease of cooling demand, total energy needs always increase with altitude for all cases. As an example, for the case of Switzerland, heating needs at 1594 m altitude are 2.4 times higher compared to those at 273 m while cooling needs are 20.5 times lower. Besides, total energy needs (heating and cooling) are still significantly higher, by 5.3 times. Presenting the results for another case study, such as Greece, it is indicated that heating needs at 980 m altitude are 1.7 times higher compared to those at 143 m (close to sea-level), cooling needs are 24.1 times lower and total energy needs are 1.3 times higher. The results show that a dwelling at a high altitude region has to spend much more money in order to achieve an adequate level of energy comfort, in all cases examined.

Moreover, the crucial role of heating energy demand is revealed. It appears that the great majority of energy needs are heating needs, not only at high altitudes but also at lower ones, with the share of heating demand versus cooling demand exceeding 70 percent at the lowest altitudes, let alone the highest ones where the share of heating needs reaches up to 99 percent.

In this way, the energy identity of an area can be determined. If the technical characteristics of buildings are also included, then an even more precise calculation of heating and cooling energy demand of buildings can be achieved, through the Eqs. (4) and (5).

4 Conclusions

Mountainous areas, fully enriched with renewable energy potential, and many times far away from national energy grids, seem to be ideal for the implementation of renewable energy technologies. Main objectives of the paper were, firstly, to overcome the obstacle of sparse meteorological stations at high altitudes and, secondly, to develop a method determining the energy demand of buildings in a simple way.

Regarding the four case studies, namely Austria, Switzerland, Greece and north Italy, four lapse rate values were estimated, at a country level. It appears that the lapse rates, indeed, differ from the common used value of 6.5°C/km, ranging from 4.9°C/km to 6.0°C/km. In this way, the air temperatures, as well as the heating degree-days of high altitude areas were calculated, using the temperature data of their nearest station. The differences arising in the calculated HDDs were found to be quite small when compared to the "real" ones, with a mean annual deviation of 7 percent, thus providing adequately reliable estimations.

Since verified that HDDs at high altitudes can be estimated with high accuracy, the energy profile of the four mountainous regions was formed. As an example, examining the energy profile along an altitudinal range of 800 m for the case of Austria, it was found that heating energy needs at the highest altitude are 1.4 times higher compared to the lowest one, cooling energy needs are 9.1 times lower and the sum of energy needs for heating and cooling are also higher, by 1.3 times. Moreover, it was proved that, in all cases, thermal energy demand is the core of energy demand, even at low altitudes.

Overall, the method suggested, based on thorough statistical analysis of long term meteorological data, provides an easy and reliable way of determining the energy needs of a mountainous region, helping in developing a tailor cut energy plan, using the abundant renewable energy sources found in mountains.

References

1. Katsoulakos, N.M., Kaliampakos, D.C.: What is the impact of altitude on energy demand? a step towards developing specialized energy policy for mountainous areas. Energy Policy **71**, 130–138 (2014)
2. Price, M.F.: Mountain Geology, Natural History and Ecosystems. Voyageur Press, Stillwater (2002)
3. Katsoulakos, N.: Optimal use of renewable energy sources in mountainous areas. The case of Metsovo, Greece (in Greek). Doctoral thesis. National Technical University of Athens, School of Mining and Metallurgical Engineering, Athens (2013)
4. Papantonis, D.: Small Hydroelectric Plants (in Greek). Symeon, Athens (2008)
5. Tsalemis, D., Mavraki, D., Doulos, H., Oikonomou, A., Perrakis, K., Tigkas, K.: Report for the power generation sector from RES, in the context of the reformation design of the support mechanism (in Greek). Technical report, MEECC, Athens (2012)
6. Bhatt, B.P., Sachan, M.S.: Firewood consumption along an altitudinal gradient in mountain villages of India. Biomass Bioenergy **27**, 69–75 (2004)
7. Büyükalaca, O., Hüsamettin, B., Yilmaz, T.: Analysis of variable-base heating and cooling degree-days for Turkey. Appl. Energy **69**, 269–283 (2001)
8. Hitchen, E.R.: Degree days in Britain. Build. Serv. Eng. Des. Tech. **2**, 73–82 (1981)
9. McMaster, G.S., Wilhelm, W.W.: Growing degree-days. One equation, two interpretations. Agric. For. Meteorol. **87**, 291–300 (1987)
10. Martinaitis, V.: Analytic calculation of degree-days for the regulated heating season. Energy Build. **28**, 185–189 (1998)
11. Kreider, J.F., Rabl, A.: Heating and Cooling of Buildings: Design for Efficiency. McGraw-Hill, New York (1994)
12. Gelegenis, J.J.: A simplified quadratic expression for the approximate estimation of heating degree days to any base temperature. Appl. Energy **86**, 1986–1994 (2009)
13. Erbs, D., Klein, S., Beckman, W.: Estimation of degree-days and ambient temperature bin data from monthly-average temperatures. ASHRAE J. **25**, 60–65 (1983)
14. Airbus: getting to grips with aircraft performance. http://www.skybrary.aero/bookshelf/books/2263.pdf
15. Technical Directive of the Technical Chamber of Greece 20701-3/2010: Climate Data of Greek Regions, 2nd edition. Athens (2012)
16. Prentice, I.C., Cramer, W.S., Harrison, P., Leemans, R., Monserud, R.A., Solomon, A.M.: A global biome model based on plant physiology and dominance, soil properties and climate. J. Biogeogr. **19**(2), 117–134 (1992)
17. Maurer, E.P., Wood, A.W., Adam, J.C., Lettenmaier, D.P., Nijssen, B.: A long-term hydrologically based dataset of land surface fluxes and states for the conterminous United States. J. Clim. **15**(22), 3237–3251 (2002)
18. Hamlet, A.F., Lettenmaier, D.P.: Production of temporally consistent gridded precipitation and temperature fields for the continental United States. J. Hydrometeorol. **6**(3), 330–336 (2005)
19. Roe, G.H., O'Neal, M.A.: The response of glaciers to intrinsic climate variability: observations and models of late Holocene variations. J. Glaciol. **55**(193), 839–854 (2010)
20. Bolstad, P.V., Swift, L., Collins, F., Regniere, J.: Measured and predicted air temperatures at basin to regional scales in the southern Appalachian mountains. Agric. For. Meteorol. **91**(3–4), 161–176 (1998)
21. Rolland, C.: Spatial and seasonal variations of air temperature lapse rates in Alpine regions. J. Clim. **16**(7), 1032–1046 (2003)

22. Blandford, T.K., Humes, K.S., Harshburger, B.J., Moore, B.C., Walden, V.P., Ye, H.: Seasonal and synoptic variations in near-surface air temperature lapse rates in a mountainous basin. J. Appl. Meteorol. Climatol. **47**(1), 249–261 (2008)

23. Minder, J.R., Mote, P.W., Lundquist, J.D.: Surface temperature lapse rates over complex terrain: lessons from the Cascade mountains. J. Geophys. Res. 115 (D14122) (2010)

24. Papada L., Kaliampakos. D.: The impact of altitude on energy demand of mountainous areas: a policy tool for Austria, Switzerland and north Italy. Manuscript Submitted for Publication (2015)

Performance Analysis of Data Mining Techniques for Improving the Accuracy of Wind Power Forecast Combination

Ceyda Er Koksoy[1], Mehmet Baris Ozkan[1]([✉]), Dilek Küçük[1], Abdullah Bestil[1], Sena Sonmez[1], Serkan Buhan[1], Turan Demirci[1], Pinar Karagoz[2], and Aysenur Birturk[2]

[1] TÜBİTAK MRC Energy Institute, Ankara, Turkey
{ceyda.er,mehmet.ozkan}@tubitak.gov.tr
[2] Middle East Technical University, 06531 Ankara, Turkey
karagoz@ceng.metu.edu.tr

Abstract. Efficient integration of renewable energy sources into the electricity grid has become one of the challenging problems in recent years. This issue is more critical especially for unstable energy sources such as wind. The focus of this work is the performance analysis of several alternative wind forecast combination models in comparison to the current forecast combination module of the wind power monitoring and forecast system of Turkey, developed within the course of the RITM project. These accuracy improvement studies are within the scope of data mining approaches, Association Rule Mining (ARM), Distance-based approach, Decision Trees and k-Nearest Neighbor (k-NN) classification algorithms and comparative results of the algorithms are presented.

Keywords: Wind power forecasting · Association rule mining · k-nearest neighbor · Decision tree

1 Introduction

Wind is an important renewable energy resource with its competitive initial costs and feasible rich potential that has a consistently growing proportion in overall energy production worldwide. However uncertainty and variability of the wind causes difficulties for transmission system operators during grid management and for wind power plant owners who need reliable information for day-ahead markets in order to maximize total return. At this point, wind power prediction systems take an important place in operational and financial usage areas such as day-ahead market, intra-day market ad real-time load balancing [1].

Similar to other Mediterranean countries such as Portugal, Spain and Italy, Turkey has also considerable wind power capacity. In Turkey there are nearly 80 Wind Power Plants (WPPs) in operation with the installed capacity of 3700 MW [2]. This number will reach 10 GW in the near future by building new WPPs whose agreements have already been signed by the government and WPP owners.

© Springer International Publishing Switzerland 2015
W.L. Woon et al. (Eds.): DARE 2015, LNAI 9518, pp. 44–55, 2015.
DOI: 10.1007/978-3-319-27430-0_4

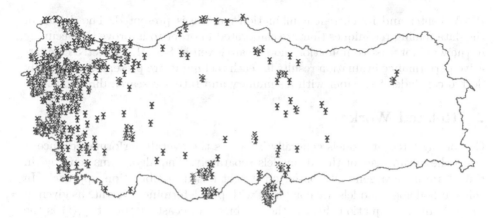

Fig. 1. Distribution of WPPs in Turkey

Generally, WPPs are located in the western part of Turkey as shown in Fig. 1. In the Eastern part of Turkey there are not too many WPPs however this region has high sun energy potential rather than wind.

As in other countries, a Wind Power Monitoring and Forecasting center (RITM)[1] has been in operation in Turkey since 2012 [2]. In this center, a statistical hybrid wind power forecast model (SHWIP) [3,4] has been applied on three different Numerical Weather Predictions (NWP) sources and at the end of the model these hourly forecasts are combined into a single final forecast with a combination algorithm. This model has been in operation since June 2012 and wind power forecasts are obtained for 51 Wind Power Plants (WPP) with total installed capacity of 2734.3 MW, which corresponds to nearly 65 % percent of the total wind capacity in the country.

The proposed SHWIP algorithm is based on clustering the weather events and finding the optimal classes. The model is applied to three different weather forecast sources, namely Turkish Meteorology Office (DMI) [9], Global Forecast System (GFS) and European Centre for Medium-Range Weather Forecasts (ECMWF), and three 48 h forecast tuples are obtained. In the next step, these forecast tuples are combined into a single 48 h forecast tuple with a combination module. Although the proposed algorithm has acceptable error rates in and it is compared with several combination methods in the literature, its combination module can be improved and model may have less error rate. In this paper several data mining approaches, namely Association Rule Mining (ARM), Distance-based approach, Decision Trees and k-Nearest Neighbor (k-NN), are elaborated on as combination module and their error rates are compared with current combination module. The methods are tested on 14 WPPs which are monitored from the beginning of project.

The rest of the paper is organized as follows: in Sect. 2, an overview of the related literature is provided and in Sect. 3, background information about

---

[1] www.ritm.gov.tr/root/index.php.

RITM center and its current combination module is presented. The details of the data mining techniques that are elaborated in order to improve the accuracy in the wind forecast combination module are given in Sect. 4. In Sect. 5 comparative performance evaluation results for each technique are presented and finally Sect. 6 concludes the paper with a summary and future research directions.

2   Related Work

Combining forecast models generally improves the overall performance since it takes the advantages of the all models separately. The algorithms reported in the literature are generally based on weight-based combination approach. In this methodology, models are combined with pre-determined weights as given in Eq. (1). In this equation, F(t) is the combined forecast at time t, $f_i(t)$ is the forecast of an individual model at the same time instance, w_i is the weight of the model and N is the total number of models used in the combination [5].

$$F(t) = \sum_{i=1}^{N} w_i f_i(t) \tag{1}$$

Revheim et al. presented a Bayes Model Averaging (BMA) method for combination [6]. They combined wind speed forecasts instead of wind power forecasts. The proposed model is based on a statistical post-processing approach in order to produce forecast from ensembles. In [7], Lyu et al. proposes a different combination strategy based on drifting degree. According to the definition, the drifting degree measures the inconsistency between the forecast value and the actual production value by examining the system deviation. According to their experimental results with their compatible and complementary model set, wind power combination forecasting model based on drifting degree is more effective in improving forecasting accuracy.

Ma et al. [8] combined their Back Propogation (BP) neural network algorithm with Genetic Algorithm (GA) and Particle Swarm Optimization (PSO) algorithm, respectively. The meteorological forecast data is given as input to BP neural network while GA and PSO are used to adjust the value of BP's connection weight and threshold dynamically. According to their experimental results, combining BP with GA produces better results compared to combining BP with PSO. In addition, they observed that the results of combination of models are much better than the results of individual models. They also compared the results of differently-weighted combination with equally-weighted combination and they concluded that the former approach performs better at determining the weights.

3   Background

3.1   Overview of RITM Center

The wind power monitoring and forecast system of Turkey is designed within the scope of the Wind Power Monitoring and Forecast Center for Turkey (RITM)

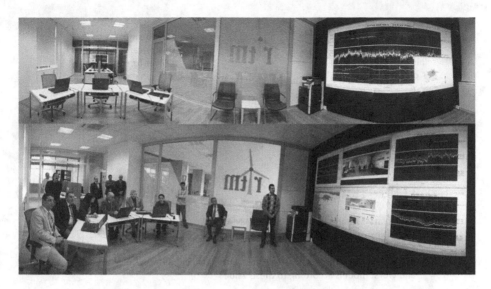

Fig. 2. A panaromic view from RITM center

project [2]. The aim of the center is monitoring the WPPs in real time and constructing a reliable forecast module in order to satisfy the economical concerns of WPP owners. In the center, the real production values are collected by the wind power analyzers settled on WPP areas in three second resolution and Numerical Weather Forecasts (NWPs) are gathered from three different sources: Turkish Meteorology Office [9], Global Forecast System (GFS) [10] and European Centre for Medium-Range Weather Forecasts (ECMWF) [11] as 48 h tuples with one hour resolution. With real production and weather forecast values, several wind power forecast models such as very short-term, short-term, regional and probabilistic model are executed daily. A panoramic view from the center is presented in Fig. 2.

3.2 Current Combination Module

Current combination of the SHWIP model is based on the combination of power forecasts rather than combination of the weather forecasts. The power forecast tuples obtained by three different NWPs are combined with a weight based approach described in Fig. 3 [3, 4]. Firstly, by using most recent 30-day of historical power data and forecast data, the average normalized mean absolute error rates for each hour are calculated independently and the forecasts of that particular hour are labelled as the best, the second and the worst forecasts. Then these forecasts are combined with their error rates as weights with cross-matching.

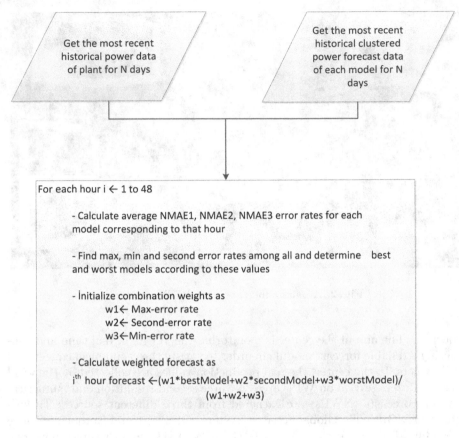

Fig. 3. Current combination module in RITM center

4   Proposed Work

In this work, four different methods are applied to wind power forecasting for constructing a combination model. The power forecasts in the RITM center are obtained from three different wind forecast sources (DMI, GFS, ECMWF) as stated in Sect. 3.1. Combination models are tested against the results of the power generation values of these NWP sources. We elaborated on the performance of data mining methods, namely association rule mining (ARM), distance-based combination, k-nearest neighbor (k-NN) and decision tree in order to produce better combined wind forecast results.

Association Rule Mining (ARM) is one of the important methods in data mining applications to find the relationship between the variables in the data set [14]. In this work, this method is adapted to wind power forecast combination problem and combination weights are determined according to support values of the forecast patterns. In the ARM-based approach, NWP training data set of WPPs is classified into six groups, which correspond to six item set patterns, with respect to forecast performance and support value of each model's performance is calculated. This value is used as the weight of NWP in the combination model. As an example, in Table 1 combination weights of the models for a sample WPP in training region is given. In the test data set, for each hourly data its corresponding class is determined and model forecasts are combined into a single final forecast according to these determined weights.

Table 1. Sample combination weights of NWPs for a sample WPP under ARM-based method

Class no	Description	Weight of NWP1	Weight of NWP2	Weight of NWP3
1	M1>M2>M3	34.85 %	24.57 %	40.57 %
2	M1>M3>M2	39.84 %	29.08 %	31.07 %
3	M2>M1>M3	25.07 %	27.05 %	47.42 %
4	M2>M3>M1	33.96 %	38.44 %	27.58 %
5	M3>M1>M2	53.55 %	15.23 %	31.21 %
6	M3>M2>M1	31.55 %	23.43 %	45.01 %

In the distance-based method, each data point is represented as a 3-dimensional vector in space, where each dimension corresponds to one of the DMI, GFS and ECMWF forecasts. Then we can calculate the distance between two data instances in the training and test data set.

K-nearest neighbor algorithm (k-NN) is a non-parametric method for classifying objects based on closest training examples in the feature space [12]. In this work we checked the performance of this approach for classifying the similar forecast tuples with different k values.

Finally, decision tree is a classification tree which maps observations about an item to conclusions about the item's class label [13]. Decision tree is considered as a feasible solution to this problem because we can set class labels of the training data set initially according to Euclidian distance to real power generation. If the algorithm can figure out the pattern of the training data set with a decision tree, we can use the extracted pattern to make decisions about the class label of test data set items.

5 Evaluation Results

This section presents performance comparison between the current combination module of RITM system and elaborated data mining based methods. The results are obtained from 14 WPPs which have been monitored in the system since the beginning of the project. Their one-year long data (six months for training and six months for test) is used in the comparison and results are based on the Normalized Mean Absolute Error (NMAE) rates whose equation is given in Eq. (2). In this formula, x_i is the real power, y_i is the estimation at i^{th} hour, C

is the installed capacity of the WPP and N is the total number of hours used in the error calculation.

$$NMAE = \frac{\sum_{i=1}^{N} \frac{|x_i - y_i|}{C}}{N} * 100. \tag{2}$$

5.1   Association Rule Mining Results

The results obtained from ARM-based method are similar to the results obtained from the current combination algorithm as shown in Table 2. The highest improvement is obtained for WPP3. Although WPP3 and WPP4 are physically neighboring plants, instead of improvement, there is deterioration by the same amount in WPP4. In 4 of 14 WPPs, the error rates are lower than that of the current combination module in these plants and WPP8 and WPP9 are the largest WPPs in the system with respect to installed capacity.

Table 2. Comparison of ARM and current combination module (in terms of NMAE)

WPP	Current combination	ARM
WPP1	**13.64** %	13.96 %
WPP2	**10.97** %	11.15 %
WPP3	16.01 %	**15.77** %
WPP4	**15.72** %	16.12 %
WPP5	**12.36** %	12.49 %
WPP6	**12.27** %	12.40 %
WPP7	**11.51** %	11.70 %
WPP8	10.78 %	**10.72** %
WPP9	11.11 %	**10.91** %
WPP10	**12.47** %	12.68 %
WPP11	**15.22** %	15.40 %
WPP12	**15.45** %	15.51 %
WPP13	**14.23** %	14.58 %
WPP14	13.39 %	**13.37%**

5.2   Distance-Based Method Results

In this approach, corresponding training data with the minimum distance, hence the maximum similarity is found for each test data. At this point different approaches can be applied. For example a threshold can be used to take into account how similar to each other date items are. As another alternative, the real power of the closest training data can be used directly as forecast.

Table 3. Comparison of distance-based and current combination module (in terms of NMAE)

WPP	Current combination	Distance-based
WPP1	**13.64** %	14.00 %
WPP2	**10.97** %	11.21 %
WPP3	16.01 %	**15.72** %
WPP4	**15.72** %	16.35 %
WPP5	**12.36** %	12.37 %
WPP6	**12.27** %	12.44 %
WPP7	**11.51** %	11.73 %
WPP8	10.78 %	**10.71** %
WPP9	11.11 %	**11.03** %
WPP10	**12.47** %	12.81 %
WPP11	**15.22** %	15.44 %
WPP12	**15.45** %	15.57 %
WPP13	**14.23** %	14.48 %
WPP14	**13.39** %	13.5 %

Our approach ignores the threshold and uses the closest training data item in order to determine weight factors of combination. The class label of the particular training data item has greater (e.g. 0.5) weight factor and the others have lower and the same weight factor (e.g. 0.25) for the particular test data item. This algorithm has been applied on each of the power plant's data and the results with the average NMAE for all of the plants are given in Table 3. In addition, the first 1-week time interval of test data set of WPP4 is presented by a line graphics in Fig. 4.

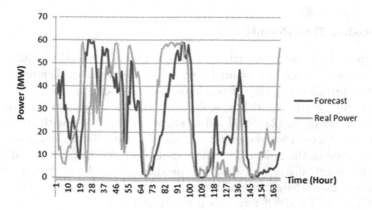

Fig. 4. Trends of partial test data of WPP4 generated by Distance-Based Method

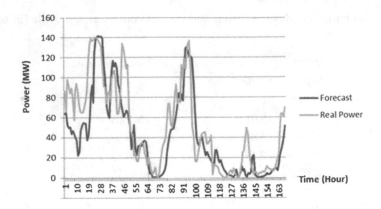

Fig. 5. Trends of partial test data of WPP10 generated by k-NN

5.3 K-Nearest Neighbour Results

Differently from distance-based method, we determine a k value and find k-nearest training data item instead of the closest one. After that, the occurrence ratio of the class labels of k-nearest training data item are used as weight factors. For example, for k = 5, algorithm determines 5-nearest neighbors. Class label is 1 (DMI) for three of them and 2 (GFS) for the rest of them. In this example, final forecast is generated from the DMI and GFS forecasts with weight factors of 0.6 and 0.4, respectively.

The results of the k-NN approach are given in Table 4. In addition, the first 1-week time interval of test data set of WPP10 is presented by a line graphics in Fig. 5. Selecting value for k is an important issue for k-NN and the results are not so stable compared to current combination algorithm. According to experimental results, in all of the WPPs the results obtained from k-NN is worse than current combination module and it shows that it is not a suitable combination approach for this data set.

5.4 Decision Tree Results

In this approach before constructing a decision tree, we need to discretize the numeric forecast values into ordinal values. We discretized all forecast values according to installed capacity of particular power plant.

In each run of the algorithm, a decision tree is constructed according to installed capacity of the plant and a discretization parameter with training data set. After that, for each test data set item, one of three different leaf labels (1, 2 or 3) is represented as the forecasting class label according to constructed decision tree. At this point, instead of using the predicted class label directly, we used the predicted class labels' probabilities as weight factors. Finally, a combination of DMI, GFS and ECMWF forecasts is generated.

Table 4. Comparison of k-NN and current combination module (in terms of NMAE)

WPP	Current combination	k-NN (k = 3)	k-NN (k = 5)
WPP1	13.64 %	17.38 %	17.38 %
WPP2	10.97 %	14.67 %	14.66 %
WPP3	16.01 %	19.20 %	19.25 %
WPP4	15.72 %	20.04 %	20.04 %
WPP5	12.36 %	14.89 %	14.90 %
WPP6	12.27 %	14.41 %	14.32 %
WPP7	11.51 %	15.12 %	15.12 %
WPP8	10.78 %	13.92 %	13.92 %
WPP9	11.11 %	13.32 %	13.31 %
WPP10	12.47 %	16.26 %	16.26 %
WPP11	15.22 %	15.41 %	15.45 %
WPP12	15.45 %	18.36 %	18.42 %
WPP13	14.23 %	15.81 %	14.45 %
WPP14	13.39 %	15.71 %	15.71 %

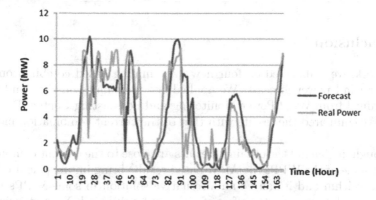

Fig. 6. Trends of partial test data of WPP6 generated by decision tree

The results of the decision tree based approach are given in Table 5. The table contains the installed capacities of the plants and also used partition (discretization) count. This information is important because, for different partition counts, different results are obtained from each run. The results obtained from this method are similar to, but mostly lower than the results of the current combination module. Only in WPP5 a small improvement is obtained and we can conclude that this method, in its current form, is not suitable for combination. To illustrate the performance of this method, the first 1-week time interval of test data set of WPP6 is presented by a line graphics in Fig. 6.

Table 5. Comparison of decision tree and current combination module (in terms of NMAE)

WPP	Current combination	Decision tree	Capacity (MW)	Partition count
WPP1	**13.64** %	13.97 %	15	3
WPP2	**10.97** %	11.15 %	90	3
WPP3	**16.01** %	16.25 %	60	5
WPP4	**15.72** %	16.35 %	35	5
WPP5	12.36 %	**12.24%**	10.2	2
WPP6	**12.27** %	12.33 %	14.9	2
WPP7	**11.51** %	12.05 %	39.2	3
WPP8	**10.78** %	11.01 %	135	5
WPP9	**11.11** %	11.21 %	140.1	7
WPP10	**12.47** %	12.57 %	36	6
WPP11	**15.22** %	15.46 %	30	3
WPP12	**15.45** %	15.75 %	30	3
WPP13	**14.23** %	14.35 %	57.5	3
WPP14	**13.39** %	13.5 %	12	4

6   Conclusion

In this work, we elaborated on four new data mining based combination techniques for wind power forecast. We applied the techniques on the wind power data obtained from Wind Power Monitoring and Forecasting Center (RITM) in Turkey. We compared the results with that of the current combination model in RITM.

Although, in general the obtained results are close to the current combination model, in some of the WPPs the ARM-based results bring improvement over the current algorithm and it can be an alternative solution in some WPPs for the future. For the future work, the effect of different k values in k-Nearest Neighbour method may be investigated. In addition, it is possible to work on a hybrid model that combines the advantages of the current model and ARM-based approach.

Acknowledgment. This work is conducted in the scope of RITM (5122807) project of TÜBİTAK. We would like thank to all of the researchers who worked in implementation of the whole project.

References

1. Hittinger, E., Apt, J., Whitacre, J.F.: The effect of variability-mitigating market rules on the operation of wind power plants. Energy Syst. **5**(4), 737–766 (2014)
2. Terciyanli, E., Demirci, T., Kucuk, D., Sarac, M., Cadirci, I., Ermis, M.: Enhanced nationwide wind-electric power monitoring and forecast system. IEEE Trans. Ind. Inf. **10**(2), 1171–1184 (2014)

3. Ozkan, M.B., Karagoz, P.: A novel wind power forecast model: statistical hybrid wind power forecast technique (SHWIP). IEEE Trans. Ind. Inf. **11**(2), 375–387 (2015)
4. Özkan, M.B., Küçük, D., Terciyanlı, E., Buhan, S., Demirci, T., Karagoz, P.: A data mining-based wind power forecasting method: results for wind power plants in turkey. In: Bellatreche, L., Mohania, M.K. (eds.) DaWaK 2013. LNCS, vol. 8057, pp. 268–276. Springer, Heidelberg (2013)
5. Tascikaraoglu, A., Uzunoglu, M.: A review of combined approaches for prediction of short-term wind speed and power. Renew. Sustain. Energy Rev. **34**, 243–254 (2014)
6. Preede Revheim, P., Beyer, H.G.: Using bayes model averaging for wind power forecasts. In: EGU General Assembly Conference Abstracts, vol. 16, p. 2811 (2014)
7. Lyu, Q.C., Liu, W.Y., Zhu, D.D., Wang, W.Z., Han, X.S., Liu, F.C.: Wind power combination forecasting model based on drift. Adv. Mater. Res. **953**, 522–528 (2014)
8. Ma, L., Li, B., Yang, Z.B., Du, J., Wang, J.: A new combination prediction model for short-term wind farm output power based on meteorological data collected by WSN. Int. J. Control Autom. **7**, 171–180 (2014)
9. Turkish Meteorology-Office Web Site. http://www.mgm.gov.tr/
10. Global Forecast System (GFS) Web Site. http://www.emc.ncep.noaa.gov/index.php?branch=GFS
11. European Centre for Medium-Range Weather Forecasts (ECMWF) Web Site. http://www.ecmwf.int
12. Hall, P., Park, B.U., Samworth, R.J.: Choice of neighbor order in nearest-neighbor classification. Ann. Stat. **36**(5), 2135–2152 (2008)
13. Rokach, L., Maimon, O.: Data Mining with Decision Trees: Theory and Applications. World Scientific Publication Co Inc. (2008). ISBN: 978-9812771711
14. Tew, C., Giraud-Carrier, C., Tanner, K., Burton, S.: Behavior-based clustering and analysis of interestingness measures for association rule mining. Data Min. Knowl. Discov. **28**(4), 1004–1045 (2014)

Evaluation of Forecasting Methods for Very Small-Scale Networks

Jean Cavallo[✉], Andrei Marinescu, Ivana Dusparic, and Siobhán Clarke

Distributed Systems Group, School of Computer Science and Statistics,
Trinity College Dublin, Dublin, Ireland
jean.cavallo@centrale-marseille.fr,
{marinesa,ivana.dusparic,siobhan.clarke}@scss.tcd.ie
http://www.dsg.cs.tcd.ie/

Abstract. Increased levels of electrification of home appliances, heating and transportation are bringing new challenges for the smart grid, as energy supply sources need to be managed more efficiently. In order to minimize production costs, reduce the impact on the environment, and optimize electricity pricing, producers need to be able to accurately estimate their customers' demand. As a result, forecasting electricity usage plays an important role in smart grids since it enables matching supply with demand, and thus minimize energy waste. Forecasting is becoming increasingly important in very small-scale power networks, also known as microgrids, as these systems should be able to operate autonomously, in islanded mode. The aim of this paper is to evaluate the efficiency of several forecasting methods in such very small networks. We evaluate artificial neural networks (ANN), wavelet neural networks (WNN), auto-regressive moving-average (ARMA), multi-regression (MR) and auto-regressive multi-regression (ARMR) on an aggregate of 30 houses, which emulates the demand of a rural isolated microgrid. Finally, we empirically show that for this problem ANN is the most efficient technique for predicting the following day's demand.

Keywords: Auto-regressive moving-average (ARMA) · Auto-regressive multi-regression (ARMR) · Artificial neural networks (ANN) · Comparison of statistical methods · Forecasting · Microgrid · Multi-regression (MR) · Smart-grid · Wavelet neural networks (WNN)

1 Introduction

As traditional power systems are transitioning towards smart grids, resilience and robustness requirements have created a demand for the decentralisation of the grid, where subcomponents of the network have to be able to operate autonomously, in islanded mode, whenever the grid requires. These subcomponents are known as microgrids, and comprise both power generation (e.g., combined-heat and power units, solar panels, wind turbines, fuel cells, diesel generators) and consumption units (e.g., factories, university campuses, households). When a microgrid is operating autonomously, it needs to be able to match

© Springer International Publishing Switzerland 2015
W.L. Woon et al. (Eds.): DARE 2015, LNAI 9518, pp. 56–75, 2015.
DOI: 10.1007/978-3-319-27430-0_5

its supply and demand, therefore requiring an estimate of future energy usage. Moreover, by combining production of electricity with heat (e.g., combined heat and power units), producers would be able to optimize their process according to both demands. Microgrids pose new challenges for electrical energy forecasting due to their very small scale. Recent research in microgrids investigate forecasting in a group of 19 customers [25], or over groups of 90/230 households [16]. Even though current microgrids are mainly implemented in campus of universities [9], they can be expanded to small towns. In order to make towns self-sufficient in terms of energy, new specific infrastructure is being built and custom forecasting tools are being developed for this kind of networks. For example, main expectations in Ireland based on 2020 European energy targets are: 40 % of electricity to be provided by renewable sources, 10 % of transport energy to be fuelled through renewable sources, 10 % penetration rate for electric vehicles and energy efficiency to be increased by 20 % [20]. However, to optimize supply and maximize integration of renewable sources, the electric energy consumption for each city has to be predicted with high reliability and accuracy. Even though some solutions for energy storage exist [2], they still do not have expected and sufficient yield. The purpose of this paper is to provide an evaluation of forecasting methods in very small scale networks, where electric energy consumption forecasting is accomplished over 50 synthetic groups of 30 houses. The demand of these groups are based on smart-meter data recorded from a set of 900 Irish houses through a trial operated by the Irish Commission of Energy Regulation in 2011. The forecasting is done on a day-ahead basis, with different statistical methods being considered.

1.1 World Electrical Consumption Repartition

The world's electric energy consumption has almost tripled[1] from 7,323 billion of KWh in 1980 to 19,710.36 in 2012. Moreover, as shown in Fig. 14, this increase is mainly due to the energy demand from developing countries in Asia. For BRIC's[2] demand, the evolution is country dependent. On the one hand, Russia's consumption stagnates thanks to their gas production. On the other hand, China has multiplied its energy usage by a factor of 17 due to strong industrial development. For developed countries, the energy use throughout the years tends to be more consistent. For example, Germany's demand has barely increased, whereas France's has doubled, and Ireland's has multiplied by a factor of three. As shown in the Fig. 1[3], the biggest recorded increases are often occurring in small countries. These are also sometimes underdeveloped countries or countries which have experienced widespread political instability. Hence, building directly efficient systems would limit the gap between them and developed countries.

---

[1] **See Fig. 14 on page xx:** http://www.eia.gov.
[2] **BRIC**: Brazil, Russia, India and China.
[3] **Source**: http://www.eia.gov and http://data.worldbank.org/.

$N°$	Country	EC (b. KWh)	Relation 2012/1980	GDP in 2012 (b. US$)
1	Bhutan	1.642	82.59	1.823
2	Maldives	0.2671	71.80	2.113
3	Turks and Caicos Islands	0.1674	36.00	nonOECD
4	Cambodia	3.004	33.65	14.054
5	Vietnam	108.27	32.96	155.820
6	Oman	20.36	25.51	77.497
7	Bangladesh	41.52	22.64	133.355
8	Cape Verde	0.29	20.47	1.757
9	Macau	4.22	18.50	42.981
10	Jordan	13.9	17.33	31.015
11	United Arab Emirates	93.28	17.10	372.313
12	China	4,467.92	17.09	9,240.270

Fig. 1. Top 12 of electrical consumption increases between 1980 and 2012

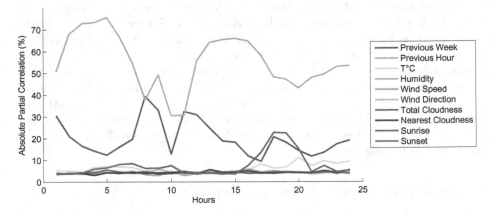

Fig. 2. Absolute partial auto-correlation function for a group of 30 houses

1.2   Indicators Impacting on Electrical Energy Consumption

Electric energy usage is dependent on several factors. Along this study, only the weather aspect is treated, whereas the social aspects of each house are outside the scope of this paper.

For each house of the forecast group, social elements impact differently on the energy demand (e.g., the social class, the size of the family and of the house, the number of rooms and the power and tariff established). However, energy demand depends also on weather conditions, and thus relies on generally available information, such as the forecast hour, the temperature along the day, the humidity, the cloudiness and the sunrise/sunset hours. Furthermore, the previous hour's consumption is impacting on future hours' demand, as shown in Fig. 2. As a side note, one of the most significant influencing factor for the overall consumption of a country is the political stability, which has a real effect on the economical stability of the country. The long term industrial and household consumptions are significantly affected by political and social situations, whereas for short term they depend on factors such as energy consumption over the previous hours/days and weather conditions.

1.3 Networks Scale and Windows Time

Most research on forecasting electrical energy usage has been done on large networks. At this scale, energy usage exhibits regular patterns. Below a certain number of houses, energy usage patterns tend to become more stochastic [12]. Such irregular patterns can be addressed through the use of artificial neural netwokrs (ANNs). ANNs are effective when used to solve non-linear problems. This is because ANNs minimize the impact of randomness on the forecasting accuracy. This is the main reason why ANNs are used in microgrid forecasting [8]. In this case, the day of the week and the month have each one two inputs. These inputs are calculated by using trigonometric functions. As shown in [8], the separation of patterns is essential when mitigating errors. As such, at least one input needs to be dedicated for the day of the week to be forecasted. The use of inputs for the month in [8] is likely due to a weak variation of temperature in Bilbao (Spain), the city where the experiments were performed. ANNs with multiple hidden layers have also been shown to be effective [7]. Nevertheless, there are some differences with the models implemented in this case: only the wind speed and the forecasted average humidity of the forecast day are considered along the input. It can be noted that the data set is not representing a microgrid but a wide area in Iran. Further work shows more focus on classical times-series forecast methods such as ARIMA, ARMA, Multi-Regression, Double Exponential Smoothing and Random Forest, where these algorithms are applied on 6 microgrids [23]. In these scenarios ARIMA is shown to be the most accurate method. There is further research where forecasting was implemented through ANNs [25], Auto-Regressive Integrated Moving-Average (ARIMA) [24,27], multi-regression (MR) [17], wavelent neural networks (WNNs) [28]. These models have been applied on different scales: large, medium, small and very small [15,25]. Finally, most of these methods forecast electricity usage on a day-ahead basis, which is classified as a short-term forecasting.

2 Background

2.1 Artificial Neural Network

A neural network comprises three type of layers: input layer, hidden layer(s) and output layer. One input corresponds to one feature and (in fully connected networks) is linked to each neuron belonging to the first hidden layer. The link weights the value of the neuron through a training process. In a neural network, there is a minimum of one hidden layer, and generally no more than three hidden layers. Each layer's neurons are linked to the previous layer's neurons. The neural network goes through a training stage, where the input sets and desired output sets are provided, and where the weights for each link auto-configure throughout the training through a process known as backpropagation. Once the network is properly trained, it is able to provide by itself outputs that closely match desired outputs. The inputs of the neural network are correlated to the outputs, as these need to give an indication of the final results of the forecast.

Remark 1. Even if there is no mathematical demonstration, the rule of thumb is for the number of neuron in a hidden layer to be between the number of inputs and outputs. Other methodology has been proposed in the literature [11,22]. According to this, when working with several hidden layers, results tend to be better when the layers have the same number of neurons.

For the forward propagation, each value of the previous layer has to be weighted and add up. Thus, Eq. (1a) allows to get the value of each hidden neurons, whereas Eq. (1b) gives the output values thanks to the same reasoning.

$$H_p = \sum_{n=1}^{N} \theta_{n,k}.I_n \tag{1a}$$

$$O_q = \sum_{p=1}^{P} \theta_{p,q}.H_p \tag{1b}$$

where θ_{ij} is the weight from $neuron_j$ of the $layer_J$ to $neuron_i$ of the $layer_I$

Equations (1a) and (1b) characterize the forward propagation. However, as every statistical method, there are also equations to correct the model during the fitting. The most used algorithm - especially because of its calculation speed - is the Resilient backPROPagation (RPROP) [10,21]. Backpropagation is very often used in supervised learning [21]. This is because it allows neural networks to adjust the weights of each neuron. This idea is characterized by the following equation:

$$\frac{\partial E}{\partial \theta_{ij}} = \frac{\partial E}{\partial s_i}.\frac{\partial s_i}{\partial net_i}.\frac{\partial net_i}{\partial \theta_{ij}} \tag{2}$$

After computing the partial derivative of an arbitrary error function E, each weight is corrected through an advanced optimization method, such as Broyden-Fletcher-Goldfarb-Shannon (BFGS) algorithm [4], or by a gradient descent:

$$\theta_{ij}(t+1) = \theta_{ij}(t) - \epsilon.\frac{\partial E}{\partial \theta_{ij}}(t) \tag{3}$$

2.2   Wavelet Neural Network

A Wavelet Neural Network (WNN) [1] is composed of two main parts [3]: the wavelet decomposition which splits the signal into several sub-signals according to high-pass and low-pass filters, and neural networks which are different for each sub-signals. Two main decomposition algorithms derive from here: Discrete Wavelet Transform (DWT) [5] and Continuous Wavelet Transform (CWT). Many algorithms can be used for DWT: Haar Wavelet, Daubechies Discrete Wavelets, Gaussian Derivative or Battle-Lemarie [26]. In this paper, only the Daubechies Discrete Wavelets (DDW) technique is presented. The common equations which have to fit function ϕ to be a wavelet [26] are: zero-mean (4a), normalization (4b) and admissibility condition (4c).

$$\int_{-\infty}^{+\infty} \phi(x)\mathrm{d}x = 0 \tag{4a}$$

$$\int_{-\infty}^{+\infty} \phi^2(x)\mathrm{d}x = 1 \tag{4b}$$

$$0 < C_\phi < +\infty, \quad C_\phi = \int_0^{+\infty} \frac{|\Phi(u)|^2}{u}\mathrm{d}u \tag{4c}$$

Each wavelet is forecasting through a separate neural network, which is calibrated during the training period. Afterwards all the results are added.

2.3 ARMA

ARMA comprises of two distinct parts. The Auto-Regressive (AR) part is a linear combination of lags of the studied time-series. The Moving Average (MA) part is a linear combination of lags of the summarized series. An ARMA model can be summarized by the process in Eq. 5. This process takes into account seasonal aspects.

$$ARMA(p,q) : y_t = \mu + \sum_{i=1}^p \phi_i.y_{t-i} - \sum_{j=0}^q \theta_j.e_{t-j} \tag{5}$$

2.4 Multi-regression

Multi-Regression attempts firstly to study the impact of external parameters which influence the signal to be predicted. If with these parameters an expert can predict the output, then the number of features is correct, otherwise, other features have to be added. After this step, the maximum polynomial degree for each feature and also the combination between each of them have to be considered. To avoid overfitting or underfitting, as described in [18], some parameters need to be computed, such as cost the function of the polynomial degree or the regularization parameter.

3 Methodology

3.1 Introduction

The data employed in the experiments is from the same database which was used in [15,16]. 900 households are grouped into 30 households sets. These 900 households comprise recorded smart-meter data from mid-July 2009 to the end 2010. 50 groups are generated in order to validate the techniques over several very small-scale networks and be able to generalize the results of the experiment. The first particularity is the common origin of all the groups. The selection of best models are done with a unique set, which represents the latest 15 weeks of the initial sampling. The final evaluation between the most efficient version of each model is done along the year of 2010 with the same groups of houses. The data used is only for the week days, as these represent a wider sample and present highest demand patterns throughout the week [16]. As the used data sets do not involve computationally demanding operations, the time difference between C++ and Matlab programs was considered irrelevant. As such, Matlab was chosen to implement all models.

3.2   Artificial Neural Network

For the implementation of this method, two solutions are possible. To build the ANN, Matlab's Neural Network Toolbox can be employed. The free open source Fast Artificial Neural Network (FANN) library [19] allows to implement ANNs in several languages such as C, PHP, Python, Octave and C++. To see how it works, a Neural Network was built without GUI. Also, a free open source GUI called FANNTool[4] is readily available for users interested to work with Neural Networks. For this forecasting method, there are two different main types of models characterized by inputs, with particular focus on the first 24 inputs used. The first model, abbreviated as ANN(W,.) is quite similar to the one described in [16]. The 31 inputs are: the previous day's hourly consumptions (24 inputs), minimum, maximum and mean of future temperature and humidity (6 inputs) and the type of the day (1 input). The only difference from the other model is due to the allocation of only one input for the type of the day for this model instead of 7. There are 24 outputs, one for each hour of the next day's electrical energy consumption, which is the forecasted information. The second variation of the model, abbreviated as ANN(D,.), uses the demand of the same day of the previous week.

3.3   Wavelet Neural Network

In this part, there are two different challenges. The first one is to find out the right Daubechies number. The second one is to use the same process described in Sect. 3.2. Based on a process inspired by methods presented in [1], the time-series represented by electrical demand is split into a subset of time-series. To ascertain the right Daubechies number, the process presented in detail in Fig. 15 was employed. This is further summarized in Fig. 3. Inputs are the same as described in [15, 16] (i.e., 24 h consumption of same day of the previous week, temperature and humidity, and day type. Outputs are the 24 h forecast of the following day).

---

[4] **Source:** https://code.google.com/p/fanntool/.

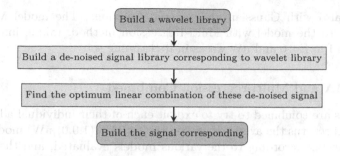

Fig. 3. Summary of process to build an optimal WNN

As described in literature [6], Daubechies number cannot be too large. The same observation is done with regard to the previous experiment. Often the right interval is between 4 and 8. That is why, the different configuration for Daubechies number are 4, 5, 6, 7 and 8. Also, for each sub-signals, such as the approximation and detail signals, the combination of wavelets characterized by their Daubechies number can be different. To simplify the reasoning, the combination has been made with final signals. The decomposition-level is 3 for all WNN models.

3.4 ARMA

Because the studied data set is already stationary, only ARMA models are set up. ARMA is built by following two different types of fitting. The first one attempts to fit the whole training set. The second type is the same as a Time-Varying Auto-Regressive Moving-Average (TVARMA), because the model's parameters are constantly updated. It works as a sliding window concept, with a window following the time frame used in the training set. Several experiments need to be carried out to estimate the optimal window-time size. The selected models for in ARMA used forecasting the energy demand comprise windows size of 4, 5, 6, 7 and 8 weeks before the testing step.

3.5 Multi-regression

For Multi-Regression, the additional parameters taken into account besides historical information are temperature and the humidity, because of their considerable impacts on electrical energy consumption. This is shown in Fig. 2.

The equation characterizing Multi-Regression is:

$$\widehat{Y}_t = \sum_{i=0}^{n} \Theta_i \cdot f\left(X_i^{(training)}\right) \tag{6}$$

The different sub-methods tested consider f as polynomial function, where the degree is ranging between a minimum of 1 to a maximum 5. Further tests were done also with Gaussian and sigmoid functions. The model MR(f,nW) corresponds to the model with Multi-Regression method, taking into account the function f in Eq. 6 and nW weeks in its training set.

3.6 ARMA and Multi-regression Combined

Both models are combined to try to exploit each of their individual advantages. This method returns the average between the (ARMA(1,0,0),nW) model, where Weeks is varying according to the various models evaluated, and the different MR(f,nW) models. This is further referred to as ARMR(f,nW).

3.7 Surrounding Forecasts Using a Frequency Approach

In this part, the aim is to evaluate the risk taken by an user by evaluating the upper and lower boundaries. For that, the training errors are assimilated to a Gaussian distribution. Then, μ and σ are calculated and both sub-signals (upper and lower), built based on $\mu + 2.\sigma$ and $\mu - 2.\sigma$, are added to the forecast signal.

4 Experiments

4.1 Introduction

For the selection of the best models for each method, one half-season has been chosen because of the high variance of consumption. Indeed, this leads to an increased level of difficulty when attempting to accurately forecast the demand. The choice of only one period for this kind of selection is due to the small size of the available data set.

4.2 Dataset Description

The initial dataset used comprises recordings from approximately 900 houses, starting from the 14th of July 2009 to the 31st of December 2010. For the purpose of evaluation, the total consumption of the 50 groups of 30 houses were considered. Thanks to the OGIMET website (www.ogimet.com), weather indicators for aviation were obtained under SYNOP code. A short function developed with Matlab has allowed to extract requested weather indicators. References for the weather station where data was measured are 03969 (WMO Index) and EIDW (ICAO Index).

Remark 2. To obtain humidity from dew-point temperature, the Heinrich Gustav Magnus-Tetens' formula was used. For some data - when temperature is under $10C$ -, $T_R > T$ - in the OGIMET file - that is implies $RH > 100\%$. To solve this kind of problem, another simple filter is set to ensure that the maximum bound is at 100%.

4.3   ANN Model Selection

Empirical tests show that the number of neurons in the hidden layer does not have a significant impact on the precision of ANNs methods with the same day in the previous week. However, both ANN(W,25) and ANN(W,26) seem to be the most accurate and reliable for this kind of models.

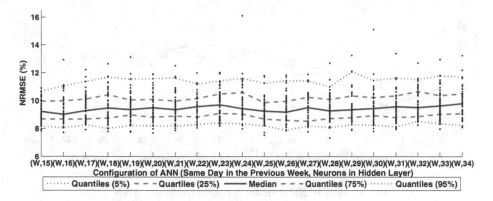

Fig. 4. Repartition of total NRMSE for each configuration - ANN - previous week

Figures 4 and 5 emphasize the efficiency of retaining electric energy consumption of the previous day instead of electric energy consumption of the same day in the previous week. Indeed, it provides on average a 5 % better median. Thus, the 5 models should be chosen among the second type of models. The contrast between accuracy and reliability for some models should be also noticed. For instance, the model ANN(D,23) is more accurate than ANN(D,20) according to the median, but it remains less reliable according to the dispersion: the 95 % quantile illustrates a difference around 1.25 % of total NRMSE.

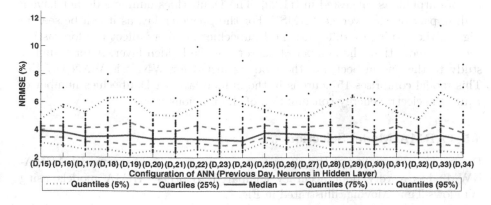

Fig. 5. Repartition of total NRMSE for each configuration - ANN - previous day

The consequences that can be drawn from evaluation are contrary to those found in [15,16] and this is in all likelihood due to size of the network studied. Actually, due to a high presence of residuals from forecasting errors, the more recent consumption is closer related to the network's future demand.

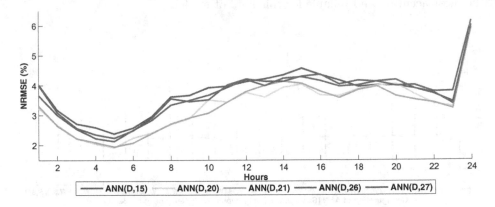

Fig. 6. Hourly NRMSE for 5 configurations - ANN

In order to refine the selection of the final method, Fig. 6 is plotted. Two methods, ANN(D,20) and ANN(D,21), have obtained best results. Finally, ANN (D,21) was selected because of its better reliability around the evening peak, which is a critical time when additional generating units can be commissioned to provide electricity. It is also because of this residual that small-scale networks forecasting is difficult.

4.4   WNN Model Selection

Figure 17 shows that for Wavelet Neural Networks, the optimal number of neurons in the hidden layer is not necessary between the number of inputs and number of outputs, as suggested in [11,22]. The Daubechies numbers do not have a high impact on the average NRMSE. For the previous day, as it can be seen in Fig. 16, the results are different: the Daubechies numbers affect the forecasting accuracy more than the number of neurons in the hidden layer. After a similar study to the one in Sect. 4.3, the model selected for WNNs is WNN(D,7,15). This model comprises 15 neurons in the hidden layer, a Daubechies number of 7 and the electric energy consumption of the previous day.

4.5   ARMA Model Selection

Due to the seasonality of the electric energy consumption, the model ARMA-NWM[5] has a very poor accuracy compared to other ARMA models using Windows-Time Moving, illustrated in Fig. 7.

---

[5] **ARMA-NWM:** ARMA No Window-Time Moving.

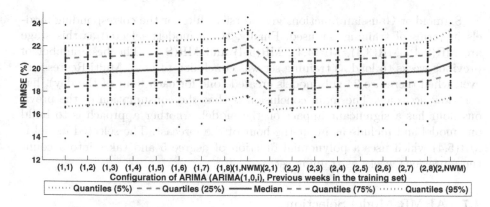

Fig. 7. Repartition of total NRMSE for each configuration - ARMA

Moreover, ARMA-WM models do not show satisfactory results, as the total NRMSE median is approximately 20 %. In this case, the 5 models which have to be selected are the first 5 models with error coefficients equal to 2. Finally, following a similar procedure to the one in Sect. 4.3, the model selected for ARMA method is (ARIMA(1,0,2),1). It can be re-written as an ARMA(1,2) model with a training set formed by only the previous week.

4.6   MR Model Selection

For the MR models, the results show that a model built without using a Window-Time Moving for the training set creates a decrease in accuracy. This decrease reaches more than 4 % of total NRMSE for the median. The second result is linked to the evolution of quartiles which gives approximately the same confidence interval, independently of the degree of the polynomial function used. It does not lead to overfitting thanks to the regularized cost function. However, it does not imply better forecasting.

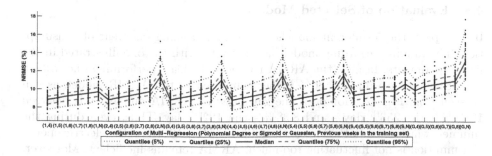

Fig. 8. Repartition of total NRMSE for each configuration - MR

Sigmoid or Gaussian functions give worst results for the corresponding models in terms of training set used. Finally, the 5 models selected as this stage are MR(2,4), MR(3,4), MR(4,4), MR(5,4) and MR(S,4). The best number of previous weeks to load as training set was concluded to be 4. Multi-Regression, even when there are 24 sub-models created (one for each hour of the day), has very low accuracy. The main conclusion is that the consumption of the previous hour has a significant impact on the model. Another approach is to build one model and include as input the hour of the forecast. The selected model is MR(5,4) which uses a polynomial function of degree 5 and takes into account the 4 previous weeks as training set.

4.7   ARMR Model Selection

The conclusions that can be drawn from Fig. 9 are similar to those from Sect. 4.6, which are shown in Fig. 8. The compensation expected between ARMA and MR models is not efficient, and leads to an overall accuracy which is between the accuracy of both models. Finally, the model selected is ARMR(5,4).

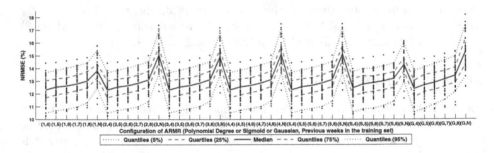

Fig. 9. Repartition of total NRMSE for each configuration - ARMR

4.8   Evaluation of Selected Models

In this part, the 5 different models are evaluated along one year of consumption. The ranking regarding models accuracy fits with the one illustrated in the first experiments. In fact, the ARMA model is the least performing one, with a NRMSE average values 12.86 % higher compared to the ARMR model, as can be seen from Fig. 12. Moreover, the gap between ARMA and ANN is between 19.45 % - the 253th day - and 8.38 % - the 122nd day. Thus, even during the summer, which is on both sides of the middle of Fig. 10, when the electric energy consumption is not fluctuating too much, ARMA remains inefficient. Moreover, it can be observed in Fig. 10 that the shape of daily NRMSE of the ARMR model and MR model are similar.

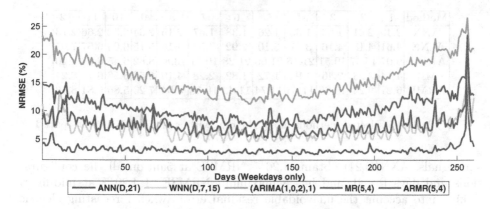

Fig. 10. Comparison of the 5 selected methods with daily NRMSE

The peak of daily NRMSE for the 256th day involves the limit of each method. Indeed, no procedure of anomaly detection of energy demand has been implemented. Thus, this day, which corresponds to Monday, 27th December 2010, is directly following Friday, 25th December 2010. It is because of only using week days in the experiments, due to their particular difference in demand patterns from weekend days. The consumption during this particular bank holiday is anomalous, and leads to overestimated inputs for the following day.

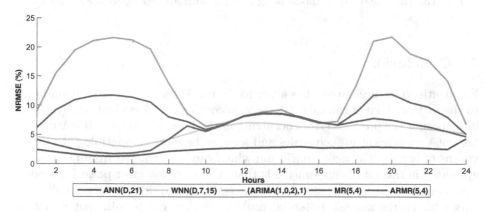

Fig. 11. Comparison of the 5 selected methods along the day with average of NRMSE

As shown in Fig. 11, from 10am to 5pm, due to the regularity of the working days, most of the models achieve the highest forecasting accuracy. Only ANN(D,21) is more robust throughout the day. The ARMA model is limited by the fluctuations of the demand during peak hours, specifically between 5–11pm, even achieving an average of 21.8 % NRMSE at 8pm.

As for WNN model, the difference of accuracy when compared to the ANN model stems from the addition of forecasting errors on the 4 predicted

Method	1	2	3	4	5	6	7	8	9	10	11	12
ANN	2.40	2.01	1.63	1.35	1.26	1.38	1.67	2.15	2.34	2.52	2.66	2.73
WNN	4.61	4.18	4.16	3.87	3.10	2.92	3.52	4.16	5.15	6.03	6.57	7.05
ARMA	9.07	15.57	19.51	21.16	21.66	21.28	19.71	13.82	8.62	6.45	6.98	8.20
MR	4.14	3.23	2.46	1.94	1.72	1.82	2.28	4.19	6.49	5.56	6.78	8.24
ARMR	6.21	9.24	10.96	11.63	11.74	11.42	10.58	8.04	7.23	5.85	6.81	8.16

Fig. 12. The 5 selected methods along the day with average of NRMSE - part 1

sub-signals. ANN(D,21) obtains 1.26 % NRMSE at 5am on all the configurations and days tested, and has a maximum NRMSE of 4.29 % at midnight. Taking into account the unavoidable residual errors when forecasting electric energy consumption in such a small-scale network, the result is considered to be sufficiently accurate for generator scheduling for this model (Fig. 13).

Method	13	14	15	16	17	18	19	20	21	22	23	24
ANN	2.78	2.83	2.94	2.74	2.75	2.81	2.87	2.83	2.78	2.66	2.50	4.29
WNN	7.03	6.79	6.35	6.25	6.23	6.69	6.58	6.62	6.19	5.96	5.57	5.18
ARMA	8.91	9.27	7.90	6.99	7.26	13.32	21.10	21.80	18.90	17.77	14.24	6.79
MR	8.71	8.63	8.11	7.18	6.65	7.27	7.80	7.49	6.85	6.34	5.53	4.67
ARMR	8.60	8.55	7.97	7.08	6.70	8.77	11.78	11.93	10.52	9.74	7.99	5.06

Fig. 13. The 5 selected methods along the day with average of NRMSE - part 2

5   Conclusion

Frequentist inference approaches allow to set the 5 % and 95 % error boundaries to ensure that the user can take a decision according to risk evaluation, by taking into account the previous errors occurring in the forecast model. ARMA models are not very efficient in such scales and seem to be more compatible with larger groups of houses. The main conclusion which can be extracted from the models evaluated in this set of experiments is that ANNs are the most powerful models that can be used to forecast electric energy consumption of very small-scale networks. Furthermore, a potential problem is that the training set is not as comprehensive as other sets used in the literature. Another method which could improve provide improvements is the use of Bayesian networks to fit the ANN, such those described in [13,14].

Acknowledgments. This work was supported by Trinity College Dublin.

A   Appendix

See the Figs. 14, 15, 16, 17 and 18.

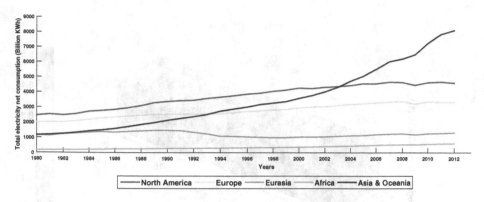

Fig. 14. Evolution of world electrical consumption by continent from 1980 to 2012

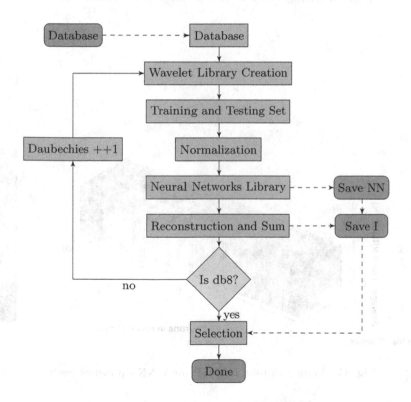

Fig. 15. Used process to build a WNN

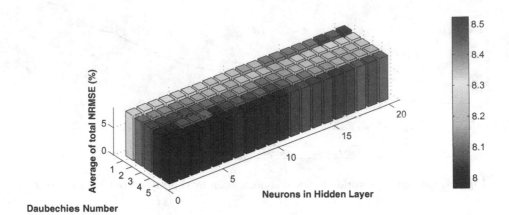

Fig. 16. Average of total NRMSE for WNN - previous day

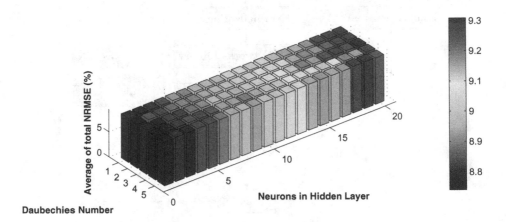

Fig. 17. Average of total NRMSE for WNN - previous week

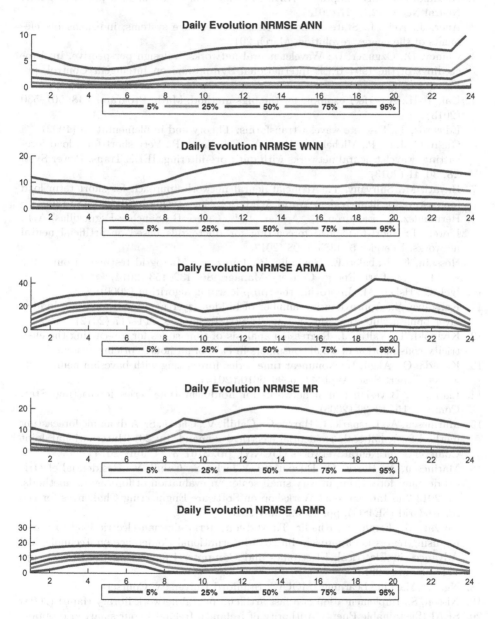

Fig. 18. Comparison of Hourly NRMSE along one year

References

1. Alexandridis, A.K., Zapranis, A.D.: Wavelet neural networks: a practical guide. Neural Netw. **42**, 1–27 (2013)
2. Auer, J., Keil, J.: State-of-the-art electricity storage systems: indispensable elements of the energy revolution, March 2012
3. Clancy, D., Ozguner, U.: Wavelet neural networks: a design perspective. In: Proceedings of the 1994 IEEE International Symposium on Intelligent Control, pp. 376–381, August 1994
4. Dai, Y.-H.: A perfect example for the bfgs method. Math. Program. **138**, 501–530 (2013)
5. Edwards, T.: Discrete wavelet transforms: Theory and implementation (1992)
6. Guan, C., Luh, P., Michel, L., Wang, Y., Friedland, P.: Very short-term load forecasting: wavelet neural networks with data pre-filtering. IEEE Trans. Power Syst. **28**, 30–41 (2013)
7. Hayati, M., Shirvany, Y.: Artificial neural network approach for short term load forecasting for illam region. World Acad. Sci. Eng. Technol. **28**, 280–284 (2007)
8. Hernandez, L., Baladrón, C., Aguiar, J.M., Carro, B., Sanchez-Esguevillas, A.J., Lloret, J.: Short-term load forecasting for microgrids based on artificial neural networks. Energies **6**, 1385–1408 (2013)
9. Hossain, E., Kabalci, E., Bayindir, R., Perez, R.: Microgrid testbeds around the world: state of art. Energy Convers. Manage. **86**, 132–153 (2014)
10. Igel, C., Hsken, M.: Improving the rprop learning algorithm (2000)
11. Karsoliya, S.: Approximating number of hidden layer neurons in multiple hidden layer bpnn architecture. Int. J. Eng. Trends Technol. **3**, 714–717 (2012)
12. Kivipõld, T., Valtin, J.: Regression analysis of time series for forecasting the electricity consumption of small consumers in case of an hourly pricing system
13. Kocada, O., Akgil, B.: Nonlinear time series forecasting with bayesian neural networks. Expert Syst. Appl. **41**, 6596–6610 (2014)
14. Liang, F.: Bayesian neural networks for nonlinear time series forecasting. Stat. Comput. **15**, 13–29 (2005)
15. Marinescu, A., Dusparic, I., Harris, C., Cahill, V., Clarke, S.: A dynamic forecasting method for small scale residential electrical demand. In: 2014 International Joint Conference on Neural Networks (IJCNN), pp. 3767–3774, July 2014
16. Marinescu, A., Harris, C., Dusparic, I., Clarke, S., Cahill, V.: Residential electrical demand forecasting in very small scale: an evaluation of forecasting methods. In: 2013 2nd International Workshop on Software Engineering Challenges for the Smart Grid (SE4SG), pp. 25–32, May 2013
17. Nezzar, M., Farah, N., Khadir, T.: Mid-long term algerian electric load forecasting using regression approach. In: 2013 International Conference on Technological Advances in Electrical, Electronics and Computer Engineering (TAEECE), pp. 121–126, May 2013
18. Ng, A.: Machine Learning (MOOC). Stanford University, Coursera
19. Nissen, S.: Implementation of a fast artificial neural network library (fann) (2003)
20. SEAI (Sustainable Energy Authority of Ireland), Ireland - your smart grid opportunity, September 2010
21. Riedmiller, M., Braun, H.: A direct adaptive method for faster backpropagation learning: the rprop algorithm. In: IEEE International Conference on Neural Networks, vol. 1, pp. 586–591 (1993)

22. Sheela, K.G., Deepa, S.N.: Review on methods to fix number of hidden neurons in neural networks. Math. Probl. Eng. **2013**, 11 (2013)
23. Subbayya, S., Jetcheva, J., Chen, W.-P.: Model selection criteria for short-term microgrid-scale electricity load forecasts. In: Innovative Smart Grid Technologies (ISGT), 2013 IEEE PES, pp. 1–6, February 2013
24. Suhartono, Puspitasari, I., Akbar, M., Lee, M.: Two-level seasonal model based on hybrid arima-anfis for forecasting short-term electricity load in indonesia. In: 2012 International Conference on Statistics in Science, Business, and Engineering (ICSSBE), pp. 1–5, September 2012
25. Tasre, M., Ghate, V., Bedekar, P.: Hourly load forecasting using artificial neural network for a small area. In: 2012 International Conference on Advances in Engineering, Science and Management (ICAESM), pp. 379–385, March 2012
26. Veitch, D.: Wavelet neural networks and their application in the study of dynamical systems, August 2005
27. Wei, L., Zhen Gang, Z.: Based on time sequence of arima model in the application of short-term electricity load forecasting. In: International Conference on Research Challenges in Computer Science, ICRCCS 2009, pp. 11–14, December 2009
28. Yang Dong, Z., Zhang, B.-L., Huang, Q.: Adaptive neural network short term load forecasting with wavelet decompositions. In: Power Tech Proceedings, 2001 IEEE Porto, vol. 2, p. 6 (2001)

Classification Cascades of Overlapping Feature Ensembles for Energy Time Series Data

Judith Neugebauer$^{(\boxtimes)}$, Oliver Kramer, and Michael Sonnenschein

Department of Computing Science, University of Oldenburg, Oldenburg, Germany
judith.neugebauer@uni-oldenburg.de

Abstract. The classification of high-dimensional time series data can be a challenging task due to the curse-of-dimensionality problem. The classification of time series is relevant in various applications, e.g., in the task of learning meta-models of feasible schedules for flexible components in the energy domain. In this paper, we introduce a classification approach that employs a cascade of classifiers based on features of overlapping time series steps. To evaluate the feasibility of the whole time series, each overlapping pattern is evaluated and the results are aggregated. We apply the approach to the problem of combined heat and power plant operation schedules and an artificial similarly structured data set. We identify conditions under which the cascade approach shows better results than a classic One-Class-SVM.

1 Introduction

In smart grids, renewable and decentralized energy units play an important role. Beside some energy units based on volatile energy resources like wind and solar power, there are also controllable power plants and consumers. The possibility to influence the operation mode of energy producers or consumers by shifting production or consumption under given constraints is called flexibility, [32]. We model these flexibilities with meta-models, which describe flexibilities as the amount of all feasible working schedules. Classifiers are appropriate meta-models with their ability to distinguish between feasible and infeasible schedules. Since schedules are time series, we chose a meta-model from time series classification. The main difference between traditional classification and time series classification tasks is, that the order of the attributes is important in time series classification tasks, [3]. For energy schedules, like Combined Heat and Power Plant (CHP) schedules, see Fig. 1(a), only the order of neighboring attributes (time steps) is important, see the autocorrelation Fig. 1(b) and the partial autocorrelation Fig. 1(c). We conduct our study on CHP and artificial data with similar properties like the CHP data set. Since all CHP schedules are further more of the same length and are not equal in shape or autocorrelation, the classification task can be treated as a traditional classification task, [25]. As we consider schedules of 24 h with 15 min resolution, the schedules have 96 attributes (time steps), this means the classification task is high-dimensional. Further more the classification task is severely imbalanced, because the class of infeasible schedules is much

© Springer International Publishing Switzerland 2015
W.L. Woon et al. (Eds.): DARE 2015, LNAI 9518, pp. 76–93, 2015.
DOI: 10.1007/978-3-319-27430-0_6

bigger in data space than the class of feasible ones, [16]. Usually imbalance is associated with samples, when there are more samples of one class than of the other class. But imbalance can also refer to the volume, that the classes occupy in space. As far as the volume of feasible CHP class in data space is only small in comparison to the much larger volume of the infeasible CHP class, it is possible to get data sets with the same number of samples from both classes, but with different distributions of the samples. For more details on the different kinds of imbalance see, [16]. In this paper imbalance refers to the volume that the classes occupy in data space.

Our preliminary experiments revealed that common classification approaches like one-class support vector machines (OCSVM), [30] are not precise enough for our application. Our analysis of the high-dimensional and imbalanced CHP time series data revealed, that comparatively simple classifiers concentrating on few time steps yield a good accuracy. Their combination to a cascade of classifiers based on patterns with overlapping features turns out to be an appropriate approach to classify the feasibility of CHP schedules. As the resulting data spaces with lower dimensionality have similar characteristics, similar parameterizations can be employed, which can reduce the effort for parameter tuning significantly.

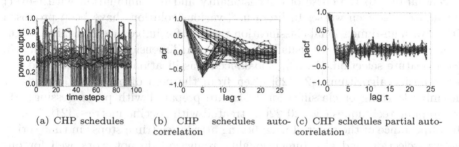

(a) CHP schedules   (b) CHP schedules auto-   (c) CHP schedules partial auto-
correlation   correlation

Fig. 1. CHP operation schedules: (a) plot of 50 normalized CHP power outputs for 24 h with 15 min resolution ($d = 96$ steps), (b) auto-correlation plotted for 20,000 schedules, (c) partial auto-correlation plotted for 20,000 schedules.

This paper is structured as follows. In Sect. 2, we give a short overview of related work on smart grid flexibilities and research on time series classification and high-dimensional and imbalanced learning. We introduce our cascade approach in Sect. 3, and present an experimental study in Sect. 4, where we compare our approach to a classic OCSVM on different data sets. We draw a conclusion in Sect. 5.

2   Related Work

During the last years, a lot of work has been conducted in the field of smart grid flexibilities and the application of optimization and learning methods. The literature differentiates between flexibility descriptions and flexibility applications. Flexibilities are mainly considered for particular applications such as single devices [8,11], while aggregated devices are employed in [28,39]. The management of flexibility descriptions is considered in [18,37]. Various methods are used to model flexibilities. Bremer *et al.* [9] employ a Support Vector Data Description meta-model to describe the flexibilities of a CHP simulation model with a thermal buffer and the flexibilities of cooling devices. We employ a similar CHP simulation model. Flexibility models based on importance weighted classification are proposed by Beygelzimer *et al.* [5].

Time series classification mainly focuses on pattern identification in time series, [15], while we are interested in the feasibility of the whole time series and their subsets, respectively. The cascade approach, we propose in this work, shares similarities with time series segmentation, where very long time series are split into shorter segments [14,15,29]. Time series segmentation is done according to time series inherent logical units, e.g. patterns, [29]. There are also similarities to feature bagging, [7,35], where the features are divided into a specific number of subsets of possibly overlapping features (feature bags). These feature bags are learned with separate models and their predictions are combined to final results.

High-dimensional classification tasks (with many attributes) are a problem in general due to the curse of dimensionality and in combination with (sever) imbalance it is even worse. In literature, various solutions have been proposed either for high-dimensional classification tasks or for imbalanced learning. High-dimensional problems are usually treated with dimensionality reduction, [13,38] or feature selection, [12,33]. Imbalanced classification tasks are learned e.g. with specific algorithms, [2,4,26] often from the field of one-class learning or ensemble learning or classification tasks are prepared with preprocessing, [19] or sampling techniques e.g. [20,22] or treated with specific metrics, [21]. Due to the importance of the order of neighboring attributes (time steps) in time series, feature selection and also dimensionality reduction do not work well for our time series. Dimensionality reduction with linear Principle Component Analysis (PCA) and PCA with a gaussian kernel revealed, that most of the components contribute to the explicable variance, see Fig. 2. The explicable variance ratio indicates the percentage of variance, that is explained by a selected component.

Most of the imbalanced classification methods are designed for lower dimensional tasks. There are only a few papers considering both, high-dimensional and imbalanced learning [6,24,27,41]. These papers focus on the classification with various facets of imbalance in combination with high dimensionality, point out the strengths and weaknesses of different methods and discuss the effect of data intrinsic characteristics. Our data sets are time series sampled from severely imbalanced classes in data space and show two specific intrinsic characteristics, recurring patterns and a correlation between neighboring attributes (time steps). Therefore, we have developed the cascade approach.

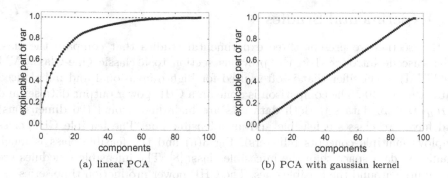

(a) linear PCA   (b) PCA with gaussian kernel

Fig. 2. Cumsum of the explicable variance ratio of 10, 000 CHP operation schedules (black line) and 100 % of the explicable variance is indicated by a red dotted line (Color figure online).

3 Cascade of Overlapping Feature Classifiers

In this section, we introduce our approach to classify time series, which are structured like CHP operation schedules. As the classification of the whole high-dimensional time series at once turned out to be difficult in preliminary experiments, we introduce a step-wise classifier based on each time series step independently. From a machine learning perspective, this approach is a cascade of classifiers based on patterns with each two overlapping features. It turns out, the single classifiers employ similarly structured data spaces and thus the effort spent on parameter tuning can be reduced.

The cascade approach works as follows. Let $(\mathbf{x}_1, y_1), (\mathbf{x}_2, y_2), \ldots, (\mathbf{x}_N, y_N)$ be a training set of N time series $\mathbf{x}_i = (x_i^1, x_i^2, \ldots, x_i^d)^T \in \mathbb{R}^d$ of d time steps, here also called dimensions (e.g. CHP schedules) with information $y_i \in \{+1, -1\}$ about their feasibility. In our approach, we train $d - 1$ classifiers $f_1, \ldots, f_{d-1}$, whose prediction will be aggregated to a final result $F(\cdot)$. Each classifier f_j is trained with the pattern-label pairs,

$$((x_1^j, x_1^{j+1}), y_1), \ldots, ((x_N^j, x_N^{j+1}), y_N). \tag{1}$$

Hence, each succeeding classifiers f_j and f_{j+1} share the features $x_1^{j+1}, \ldots, x_N^{j+1}$, which is important to evaluate the feasibility of each single time step. The whole time series is feasible, if all classifiers in the cascade confirm the feasibility of each step:

$$F(\mathbf{x}) = \begin{cases} +1 & \text{if } f_i \neq -1 \; \forall i = 1, \ldots, d - 1 \\ -1 & \text{else} \end{cases} \tag{2}$$

This approach can be extended in various kinds of ways, e.g., to longer time series intervals, longer overlaps, and w.r.t. the kind of aggregation of the classifier results, e.g., a majority vote might be useful in certain applications.

4 Experimental Studies

In this section we present three experimental studies that compare the classifier cascade introduced in the previous section to a classic One-Class SVM (OCSVM), a classifier that is often used for high dimensional and imbalanced data, see e.g. [40]. The comparison is done on a CHP power output data set and a *Hypersphere* data set. Both data sets are high-dimensional (96 dimensions) and have two classes a feasible and an infeasible one. The feasible CHP class employs multiple concepts (clusters), Fig. 3(a) and the feasible class is much smaller in data space than the infeasible class, [8]. The infeasible schedules are distributed around the feasible ones. The CHP power production time series are simulated with a CHP simulation model[1] including a thermal buffer and the heat demand of one household. The experiments are conducted with normalized CHP schedules, where the power output is normalized to values between 0 and 1. The feasible class of the *Hypersphere* data set consists of four hypersphere concepts (clusters), Fig. 3(b). The infeasible class occupies the remaining volume. But infeasible examples are sampled only in "rings" around the hyperspheres near the boundaries, for testing the selectivity of the learned classifiers. These four hypersphere concepts are generated by stretching and translating hypersphere data simulated with Dd_tools, [36]. To adapt the cascade approach optimally to the data sets, different baseline methods are employed on the small, lower dimensional data sets. In our study, we employ the cascade approach with the OCSVM from SCIKIT-LEARN, [30], a density based method from Dd_tools [36], a mixture of Gaussians (MOG) and a clustering based method kmeans data description (kmeans) from Dd_tools [36]. The classic OCSVM for comparison is from SCIKIT-LEARN, [30].

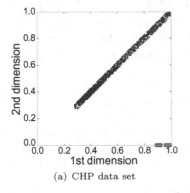
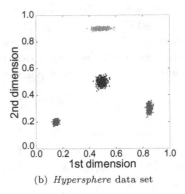

(a) CHP data set   (b) *Hypersphere* data set

Fig. 3. The first 2 dimensions of the feasible classes of the 96-dimensional data sets.

The three experimental studies have an increasing complexity. The first study (Study A) in Sect. 4.1 compares the cascade approach and a classic OCSVM on

---

[1] Data are available for download on our department website http://www. uni-oldenburg.de/informatik/ui/forschung/themen/cascade/.

both data sets, where the feasible class is reduced to one concept (cluster). The second study (Study B) in Sect. 4.2 repeats the experiments from Sect. 4.1 on the complete data sets. Therefore the cascade approach is adapted to all concepts of the feasible class. In a further analysis the impact of the 2-dimensional classifiers in the cascade is analyzed and the cascade approach is extended to arbitrary dimensional small classifiers in Study C in Sect. 4.3. The studies are evaluated according to true positive (TP) rates and true negative (TN) rates. In literature more advanced measures, e.g. see [16,21] are available for different kinds of classification. But the three studies are conducted with feasible example, some also with infeasible examples or with infeasible examples only near the class boundaries. For this reason the simple TP- and TN rates yield the best comparability among the studies.

4.1 Experimental Study A

In this section, we compare the classifier cascade introduced in the previous section to a classic OCSVM on the reduced data sets. The feasible class of the CHP data set is reduced to CHP schedules, where the power production is always greater than 0 (CHP is switched on in all 96 time steps), see the large blue cluster in Fig. 3(a). The feasible class of the *Hypersphere* data set is reduced to the blue hypersphere in the middle of Fig. 3(b) and for the experiments only infeasible schedules around the selected middle hypersphere are used.

Experimental Settings. The performance of the cascade approach is analyzed in comparison to a classic OCSVM. The cascade approach is employed with different baseline classifiers OCSVM, MOG and kmeans. The comparison is conducted for varying numbers N of feasible training examples $N = 1000, 2000, \ldots, 15000$ (CHP) and $N = 1000, 2000, \ldots, 10000$ (*Hypersphere*). The classifier parameters are optimized with grid-search on a separate independent validation sets of the same size as the training set. Parameter optimization of each 2-dimensional classifier from the cascade approach is done according to true positive rates (TP rate or only TP), ($TP\ rate = (true\ positives)/(number\ of\ feasible\ examples)$), because there are no 2-dimensional infeasible examples available. The high-dimensional infeasible examples are infeasible only in high-dimensional space and their 2-dimensional projections can be identical to 2-dimensional projections of feasible examples. For better comparability, the parameters of the classic OCSVM are also optimized with grid-search according to TP rates. The OCSVM parameters for the classic OCSVM and the cascade approach with OCSVM are optimized in defined ranges for γ^2 and ν^3. The MOG parameters are optimized in specific ranges, which turned out to be appropriate in pre-studies, for error ϵ^4 on the target class and the number $k_t{}^5$ of clusters for the target data and with setting $k_o = 0$ as outlier parameter. The kmeans parameters are optimized in the

---

[2] $\gamma \in \{0.1, 1, 10, 50, 100, 150, 200\}$.
[3] $\nu \in \{0.0001, 0.001, 0.0025, 0.005, 0.0075, 0.01, 0.025, 0.05, 0.075, 0.1, 0.2\}$.
[4] $\epsilon \in \{0.01, 0.05, 0.1, 0.15, 0.2\}$.
[5] $k_t \in \{1, 5, 10, 15\}$.

following ranges of nearest neighbors $k \in \{1, 2, \ldots, 20\}$ and target error values $\epsilon_t \in \{0.01, 0.05, 0.1, 0.15\}$. The final classifiers for each data set are tested on the same independent data set with 10,000 feasible examples and 10,000 infeasible examples. The CHP data contains infeasible examples randomly sampled from the region of infeasible schedules. The *Hypersphere* data set contains infeasible examples, that are distributed only near the class boundary around the *Hypersphere*, for testing the selectivity of the learned classifiers.

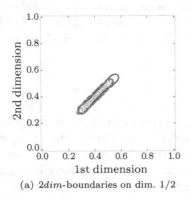

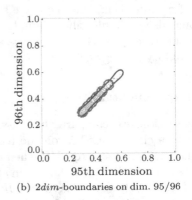

(a) *2dim*-boundaries on dim. 1/2   (b) *2dim*-boundaries on dim. 95/96

Fig. 4. Decision boundaries on the reduced CHP data, OCSVM (solid blue), kmeans (solid olive) and MOG (dashed red). Gray scattered points indicate 500 of $N = 5000$ training examples (Color figure online)

Results. Figure 4 shows the learned boundaries of the first, Fig. 4(a) and the last classifier of the cascade, Fig. 4(b) from the cascade approach for the CHP data set. The same plots for the *Hypersphere* data set are shown in Fig. 6(a) and (b). Especially for the 2-dimensional CHP data set the 2-dimensional OCSVM classifiers learn larger regions of feasible schedules than the MOG and the kmeans classifiers and as a result the OCSVM cascade predicts more feasible test schedules as feasible, see Figs. 5(a) and 7(a). The classic OCSVM achieves the highest TP rates. For the CHP data set all classifiers achieve for all values of N TN rates of 1, see Fig. 5(b), because most of the infeasible examples are located far away from the class boundary. All in all the classic OCSVM shows the highest TP rates of all classifiers and also high TN rates. For the *Hypersphere* data set, where the infeasible test examples are located near the class boundary, the OCSVM cascade has the lowest TN rate and the classic OCSVM the highest, see Fig. 7(b). Increasing values of N lead to increasing TP rates and decreasing TN rates for the cascade approach, see Fig. 7. The more feasible examples the classifier predicts as feasible, the wider is the learned class boundary and the more infeasible examples near the class boundary lie inside the learned region of the feasible class. On both data sets the classic OCSVM achieves higher TP rates and on the *Hypersphere* data set even higher TN rates than our cascade

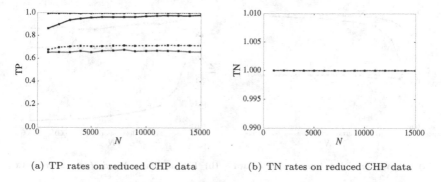

(a) TP rates on reduced CHP data   (b) TN rates on reduced CHP data

Fig. 5. TP and TN rates of the classification on the reduced CHP data set. The lines indicate cascade classification with OCSVM (solid blue), kmeans (solid olive), MOG (dashed red) and classic OCSVM (dashed dotted black). All TN rates have a value of 1 (Color figure online).

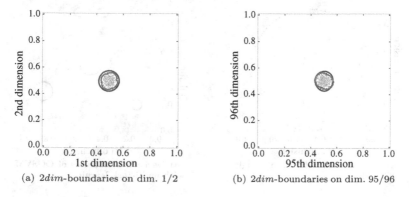

(a) $2dim$-boundaries on dim. 1/2   (b) $2dim$-boundaries on dim. 95/96

Fig. 6. Decision boundaries on the reduced *Hypersphere* data, OCSVM (solid blue), kmeans (solid olive) and MOG (dashed red). Gray scattered points indicate 500 of $N = 5000$ training examples (Color figure online)

approach. For these two reduced data sets with only one concept the cascade approach can not keep up with the classic OCSVM.

4.2 Experimental Study B

Feasible classes of time series classification tasks can consists of more than one (simple) cluster as in Sect. 4.1. In this section we consider the complete data sets, where the feasible class consists of several concepts (clusters). Learning several clusters with a cascade of 2-dimensional classifiers leads to imprecise 2-dimensional classifiers, Fig. 8(a). Yang *et al.* proposed multi-task learning in [40] for data sets where one class consists of more than one concept. Multi-task learning means learning each concept (cluster) separately and merging the classification results to an overall result, Fig. 8(b). Multi-task learning can also be applied

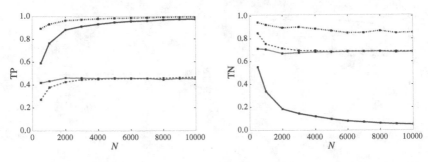

(a) TP rates on reduced *Hypersphere* data   (b) TN rates on reduced *Hypersphere* data

Fig. 7. TP and TN rates of the classification on the reduced *Hypersphere* data set. The lines indicate cascade classification with OCSVM (solid blue), kmeans (solid olive), MOG (dashed red) and classic OCSVM (dashed dotted black) (Color figure online).

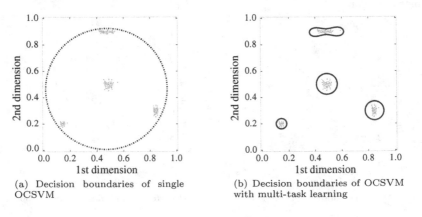

(a) Decision boundaries of single OCSVM   (b) Decision boundaries of OCSVM with multi-task learning

Fig. 8. OCSVM boundaries on the first two dimensions of the *Hypersphere* data set.

to high-dimensional data, [17,31,34] or to time series classification, [1]. For overlapping and inhomogeneously sampled concepts, clustering analysis, [23] could be also applied instead of multi-task learning. For the experimental study B we extend our cascade approach by multi-task learning. Depending on the data set and how the concepts are separable in high and in low-dimensional space, we propose two ways to combine our cascade approach with multi-task learning.

1. First case: The concepts are better separable in low-dimensional space. In this case the data set is split into 2-dimensional data subsets as before, see (1). These 2-dimensional data subsets are split according to the concepts (clusters) based on knowledge about the concepts or based on clustering. In contrast to the basic cascade approach in Sect. 3 where one classifier is trained per cascade step, now one classifier is trained for each concept per cascade step. The results of the different classifiers on the test set are combined as before, see (2).

2. Second case: The concepts are better separable in high-dimensional space. The high-dimensional data set is split into subsets according to the high-dimensional concepts. Each concept is learned with the cascade approach as a single data set in Sect. 3. The classification yields results for each concept. To receive an overall result, the Confusion matrices of all concepts are calculated and aggregated.

Experimental Settings. We apply the same baseline classifiers for the cascade approach and the same classic OCSVM as in Sect. 4.1 and extend the list of methods by a min-max classifier. This min-max classifier is designed for the 2-dimensional CHP concepts at the edges with $x = 0$ and or $y = 0$. The classifier learns whether there is a feasible example at $(0,0)$ and it learns the minimum >0 and the maximum for the x and the y axis, see Fig. 9. New examples are classified as feasible, when they are located on the x or y axis between the minimum and the maximum or when they are located at $(0,0)$ and this was a feasible training example. (Due to some CHP hard constrains concerning the CHP operation mode, there can be no feasible examples between 0 and $min > 0$, [8].) The classifiers for the CHP data set (classic OCSVM without multi-task learning, cascade with multi-task learning OCSVM + min-max, MOG + min-max, kmeans + min-max and cascade with OCSVM without multi-task learning) are learned with N training examples from both concepts (clusters) and with multi-task learning (first case). The *Hypersphere* data set is classified (classic OCSVM with and without multi-task learning and the cascade approach with multi-task learning with OCSVM, MOG and kmeans for each concept) with N examples of each concept and multi-task learning according to the second case. Parameter optimization is done as in Sect. 4.1. The CHP classifiers are tested on a test set with $10,000$ feasible examples of both concepts and $10,000$ infeasible examples. The *Hypersphere* classifiers are tested on a test set with $10,000$ feasible

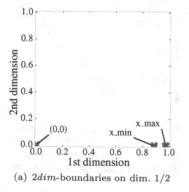

(a) *2dim*-boundaries on dim. 1/2

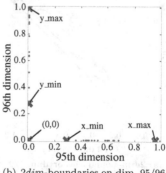

(b) *2dim*-boundaries on dim. 95/96

Fig. 9. CHP concepts on the axes learned with the min-max classifier. Red crosses mark the minimum (>0) and maximum for both axes as well as the origin $(0,0)$, which can be feasible. The blue points indicate training examples (Color figure online)

examples of each concept and 10,000 infeasible examples near the boundary of each concept.

Results. The learned class boundaries of the 2-dimensional classifiers on both data sets are more precise with multi-task learning, while the class boundaries without multi-task learning comprise more infeasible examples, see Figs. 10 and 12. The OCSVM without multi-task learning and the OCSVM with multi-task learning (OCSVM + min-max) achieve similar decision boundaries for the concept in the middle (black line is hidden by the solid blue line), but the decision boundaries for the concepts on the edges are different (black line and blue crosses). But the overall classification result of the 96-dimensional CHP data set shows the highest TP rates for the classic OCSVM and the OCSVM cascade without multi-task learning, Fig. 11(a). For higher numbers of training examples N, the TP rate of the OCSVM cascade with multi-task learning approaches the TP rate of the classic OCSVM and the TP rate of the OCSVM cascade without multi-task learning. The TN rate of the CHP classification is near 1 for all classifiers, except for the classic OCSVM where the TN rate decreases for increasing values of N, Fig. 11(b). The classic OCSVM achieves the highest TP rates because the classifier overestimates the feasible regions at the cost of misclassified infeasible examples. The fact, that the infeasible test examples are not located near the class boundary shows, the classic OCSVM overestimates the feasible class to a great degree. Therefore the OCSVM cascade with multi-task learning is a very good classifier for larger values of N, even though the feasible class is a bit underestimated.

The classification results of the *Hypersphere* data set in Fig. 13 are similar to the results of the *Hypersphere* data set in Sect. 4.1 concerning the TP rates. The classic OCSVM and the classic OCSVM with multi-task learning achieve the highest TP rates, Fig. 13(a). But due to the overestimation of the feasible

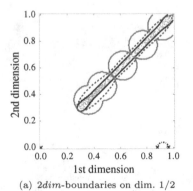
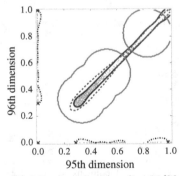

(a) *2dim*-boundaries on dim. 1/2	(b) *2dim*-boundaries on dim. 95/96

Fig. 10. Decision boundaries on CHP, concept in the middle with OCSVM (solid blue), kmeans (solid olive) and with MOG (dashed red), crosses mark min-max boundaries and OCSVM without multi-task learning (dashed dotted black). The gray and cyan points indicated 500 of $N = 5000$ training examples (Color figure online).

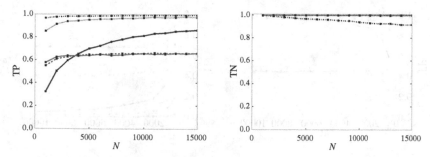

(a) TP rates of CHP data with multi-task learning   (b) TN rates of CHP data with multi-task learning

Fig. 11. TP and TN rates of the 96-dim CHP with classic OCSVM (dashed dotted black) and cascade approach: OCSVM + min-max (solid blue), kmeans + min-max (solid olive), MOG + min-max (dashed red) and a single OCSVM (dotted green) (Color figure online).

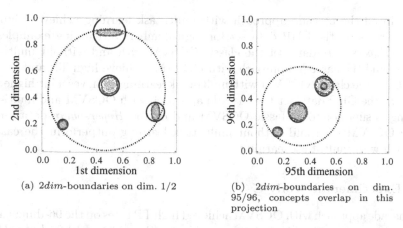

(a) 2*dim*-boundaries on dim. 1/2

(b) 2*dim*-boundaries on dim. 95/96, concepts overlap in this projection

Fig. 12. Decision boundaries on *Hypersphere*, with multi-task learning OCSVM (solid blue), kmeans (solid olive), MOG (dashed red) and OCSVM without multi-task l. (dashed and dotted black). The gray points indicate 500 of $N = 5000$ training examples (Color figure online).

class by the classic OCSVM, most of the infeasible training examples are misclassified. This leads to low TN rates for the classic OCSVM, Fig. 13(b). The classic OCSVM achieves higher TN rates with multi-task learning. The cascade approach with OCSVM, MOG and kmeans classifiers achieve a bit higher TN rates than the classic OCSVM at the cost of lower TP rates. The classifiers on the *Hypersphere* data set show increasing TP rates for increasing values of N and for high values of N the TP rates approach similar values, except for the MOG and the kmeans classifier, which are therefore not of interest. The TN rates of all classifiers decrease with increasing values of N.

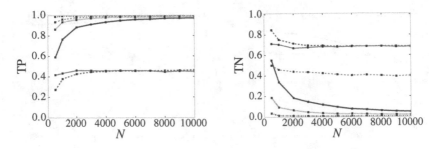

(a) TP rates of *Hypersphere* data with multi-task learning   (b) TN rates of *Hypersphere* data with multi-task learning

Fig. 13. TP and TN rates of the 96-dim *Hypersphere* with classic OCSVM without multi-task learning (dashed dotted black), with (dashed cyan) and cascade approach with multi-task learning: OCSVM (solid blue), kmeans (solid olive) and MOG (dashed red) (Color figure online).

All in all the cascade approach with multi-task learning achieves high TP and TN rates on the CHP data set for high numbers of training examples N. On the *Hypersphere* data set the classic OCSVM with and without multi-task learning and the cascade approach with OCSVMs achieve high TP and low TN rates. Only the classic OCSVM with multi-task learning achieves a bit higher TN rates. On the CHP data set the cascade approach with OCSVM and multi-task learning is superior to a classic OCSVM and on the *Hypersphere* data set the classic OCSVM with and without multi-task learning outperform the cascade approach with multi-task learning.

4.3   Experimental Study C

The cascade approach with OCSVM achieved high TP rates on the 96-dimensional *Hypersphere* data set in the previous experiments (Sects. 4.1 and 4.2), but only low TN rates for the infeasible *Hypersphere* test examples near the class boundaries. The previous experiments were based on 2-dimensional-cascades, but for some data sets, e.g., the *Hypersphere* data set, the projections to lower dimensional space lead to a loss of information. For example, a $3d$ sphere learned with a 2-dimensional OCSVM cascade classifier leads to a larger volume of the feasible class than the volume of the original sphere, see Fig. 14. Our pre-investigations have shown that the over-estimation of the sphere with a 2-dimensional classifier increases with increasing data set dimensionality. In this section, we extend the cascade approach from 2-dimensional-cascades, to cascades of arbitrary dimensionality p. The data set is divided into p-dimensional data subsets with $p < d$. In comparison to the 2-dimensional-cascade, we train $d - (p-1)$ classifiers $f_1, \ldots, f_{d-(p-1)}$, whose predictions are again aggregated to a final result $F(\cdot)$. Each classifier f_j is trained with the pattern-label pairs,

$$((x_1^j, \ \ldots, \ x_1^{j+p-1}), y_1), \ldots, ((x_N^j, \ \ldots, \ x_N^{j+p-1}), y_N), \ j = 1, \ldots, d-1. \quad (3)$$

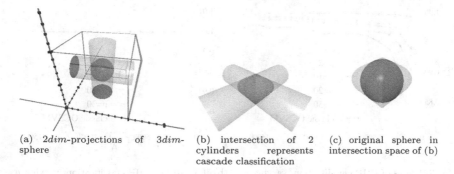

(a) 2*dim*-projections of 3*dim*-sphere

(b) intersection of 2 cylinders represents cascade classification

(c) original sphere in intersection space of (b)

Fig. 14. Geometric example of a 2-dimensional-cascade classification of a $3d$ sphere. This example shows a systematic classification (projection) error. The figures are plotted with POV-Ray, [10].

Depending on the classification task two neighboring low-dimensional data sets can overlap within a number of features between 1 and $p-1$. In our experiments, we employ an overlap of $p-1$ features.

Experimental Settings. The experiments on the CHP data set are conducted with the same settings as in Sect. 4.1 for schedules with power production >0. Instead of the 2-dimensional-cascade classifiers, we use p-dimensional classifiers with $p \in \{2, 10, 20, 30\}$ and only OCSVM baseline classifiers. The experiments on the *Hypersphere* data set are done with the settings of Sect. 4.2 with multi-task learning and p-dimensional cascade classifiers with $p \in \{2, 20, 40, 60, 80, 96\}$ and only OCSVM baseline classifiers.

Results. The classification of the CHP middle concept shows increasing TP rates for increasing values of p, see Fig. 15(a) and TN rates of about 1 for all considered classifiers, see Fig. 15(b). The classic OCSVM with $p = 96$ achieves the highest TP rate on the CHP middle concepts. On the *Hypersphere* data set, the TP rates of all p values increase with increasing values of N. For small values of N, Fig. 16(a) shows also increasing TP rates for increasing values of p. The higher the value of p, the slower converge the TP rates. This leads to intersections of the TP rate lines. The classic OCSVM achieves the highest TP rates and at the same time the lowest TN rates. Varying values of p show a different behavior for the TN rates than for the TP rates. The higher the value of p the higher is the TN rate for all values of N, Fig. 16(b). Increasing values of N lead to decreasing TN rates. In summary, the cascade approach with varying values p yields high TP and relatively high TN rates when p is optimized with respect to the data set and the number of training examples N. On the reduced CHP data set the classic OCSVM achieves a bit better results than the cascade approach with varying values p. But on the *Hypersphere* data set the cascade approach with varying values p and a careful adaptation is superior to a classic OCSVM.

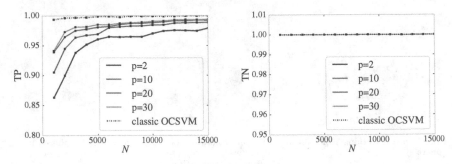

(a) TP rates of CHP classification, varying p (b) TN rates of CHP classification, varying p

Fig. 15. TP and TN rates of the 96-dimensional reduced CHP data set. Results are indicated for OCSVM with p-dimensional cascades, resp. classic OCSVM.

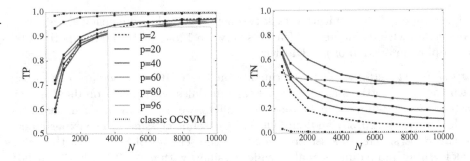

(a) TP rates of *Hypersphere* classification with   (b) TN rates of *Hypersphere* classification with
multi-task learning and varying values of p multi-task learning and varying values of p

Fig. 16. TP and TN rates of the 96-dimensional *Hypersphere* data set. Results are indicated for OCSVM with p-dimensional cascades, resp. classic OCSVM. Legend of Fig. 16(a) is also valid for Fig. 16(b)

5   Conclusions

The proposed cascade approach based on simple classifiers with overlapping features showed promising results on the simulated CHP power output data set and the artificial *Hypersphere* data set. In three experimental studies, we have compared the performance of our cascade approach with OCSVM, kmenas and MOG baseline classifiers to a classic OCSVM. The cascade approach yields better results on the complex CHP and *Hypersphere* data sets with multiple concepts, see Fig. 13, while a classic OCSVM performs better on the reduced CHP data sets where the feasible class consists only of one simple concept. The cascade approach performs much better with OCSVMs as baseline classifiers on both data sets than with the density based MOG classifiers or the clustering based kmeans classifiers, because the low dimensional data sets are not homogeneously distributed in the region of feasible schedules. In low-dimensional space of the cascade classifiers the curse of dimensionality is no longer a problem, but projection errors can occur for the *Hypersphere* data set. A careful choice of the dimensionality of the low-dimensional data sets can solve this problem, as shown in Fig. 16 for the *Hypersphere* data set. The analysis of the decision boundaries of the low-dimensional cascade classifiers on the two data sets revealed, that carefully chosen and adapted base classifiers lead to more precise class boundaries than a classic OCSVM. These findings were confirmed by optimally adapted cascade classifiers to the *Hypersphere* data set. They achieved high TP rates and also high TN rates on the infeasible examples near the class boundaries, see Fig. 16. This means in particular that the cascade approach classifies less infeasible examples as feasible, because the learned class boundaries surround the feasible class closer than the boundaries learned by a classic OCSVM. This is important for smart grid applications with prediction and optimization tasks. Here, a complete set of all feasible schedules is often required and infeasible schedules predicted as feasible are worse than feasible schedules predicted as infeasible. In summary, it can be said that the cascade approach can be used as a flexibility meta-model, i.e., a description of all feasible schedules, that performs well on the complex high-dimensional CHP and the *Hypersphere* data set with a careful adaptation to the data sets.

In future work, we plan to analyze the impact of data preprocessing on the cascade approach and to extend the approach to larger aggregations of flexible actors in smart grids.

Acknowledgments. This work was funded by the Ministry for Science and Culture of Lower Saxony with the PhD program System Integration of Renewable Energy (SEE).

References

1. Al-Hmouz, R., Pedrycz, W., Balamash, A., Morfeq, A.: Description and classification of granular time series. Soft. Comput. **19**(4), 1003–1017 (2015)
2. Attenberg, J., Ertekin, C.: Class Imbalance and Active Learning. Wiley, Hoboken (2013)
3. Bagnall, A., Davis, L.M., Hills, J., Lines, J.: Transformation based ensembles for time series classification. In: Proceedings of the Twelfth SIAM International Conference on Data Mining, Anaheim, California, USA, pp. 307–318, 26–28 April 2012
4. Batuwita, R., Palade, V.: Class Imbalance Learning Methods for Support Vector Machines. Wiley, New York (2013)
5. Beygelzimer, A., Dani, V., Hayes, T., Langford, J., Zadrozny, B.: Error limiting reductions between classification tasks. In: Proceedings of the 22nd International Conference on Machine Learning, ICML 2005, pp. 49–56. ACM, New York (2005)
6. Blagus, R., Lusa, L.: Class prediction for high-dimensional class-imbalanced data. BMC Bioinform. **11**(1), 523 (2010)
7. Breiman, L.: Random forests. Mach. Learn. **45**(1), 5–32 (2001)
8. Bremer, J., Rapp, B., Sonnenschein, M.: Support vector based encoding of distributed energy resources' feasible load spaces. In: Innovative Smart Grid Technologies Conference Europe IEEE PES (2010)

9. Bremer, J., Sonnenschein, M.: Model-based integration of constrained search spaces into distributed planning of active power provision. Comput. Sci. Inf. Syst. **10**(4), 1823–1854 (2013)
10. Buck, D.K., Collins, A.A.: POV-Ray - The Persistence of Vision Raytracer (2004). Computer software available from http://www.povray.org/
11. Castillo-Cagigal, M., Martín, E.C., Matallanas, E., Masa-Bote, D., Gutiérrez, A., Monasterio-Huelin, F., Jiménez-Leube, J.: PV self-consumption optimization with storage and active DSM for the residential sector. Sol. Energ. **85**(9), 2338–2348 (2011)
12. Chandrashekar, G., Sahin, F.: A survey on feature selection methods. Comput. Electr. Eng. **40**(1), 16–28 (2014). 40th-year commemorative issue
13. Engel, D., Hüttenberger, L., Hamann, B.: A survey of dimension reduction methods for high-dimensional data analysis and visualization. In: Garth, C., Middel, A., Hagen, H. (eds.) Visualization of Large and Unstructured Data Sets: Applications in Geospatial Planning. Modeling and Engineering - Proceedings of IRTG 1131 Workshop 2011, volume 27 of OpenAccess Series in Informatics (OASIcs), pp. 135–149. Schloss Dagstuhl-Leibniz-Zentrum fuer Informatik, Dagstuhl, Germany (2012)
14. Fulcher, B., Jones, N.: Highly comparative feature-based time-series classification. IEEE Trans. Knowl. Data Eng. **26**(12), 3026–3037 (2014)
15. Fulcher, B.D., Little, M.A., Jones, N.S.: Highly comparative time-series analysis: the empirical structure of time series and their methods. J. R. Soc. Interface, **10**(83) (2013)
16. He, H., Garcia, E.: Learning from imbalanced data. IEEE Trans. Knowl. Data Eng. **21**(9), 1263–1284 (2009)
17. He, X., Mourot, G., Maquin, D., Ragot, J., Beauseroy, P., Smolarz, A., Grall-Maës, E.: Multi-task learning with one-class SVM. Neurocomputing **133**, 416–426 (2014)
18. Hinrichs, C., Sonnenschein, M., Lehnhoff, S.: Evaluation of a self-organizing heuristic for interdependent distributed search spaces vol. 1 - agents. In: Filipe, J., Fred, A.L.N. (eds.) International Conference on Agents and Artificial Intelligence (ICAART 2013), pp. 25–34. SciTePress, Germany (2013)
19. Hoens, T.R., Chawla, N.V.: Imbalanced Datasets: From Sampling to Classifiers. Wiley, New Jersey (2013)
20. Hwang, W., Runger, G., Tuv, E.: Multivariate statistical process control with artificial contrasts. IIE Trans. **39**(6), 659–669 (2007)
21. Japkowicz, N.: Assessment Metrics for Imbalanced Learning. Wiley, New Jersey (2013)
22. Juszczak, P., Duin, R.P.W.: Selective sampling methods in one-class classification problems. In: ICANN, pp. 140–148 (2003)
23. Kang, J.H., Kim, S.B.: A clustering algorithm-based control chart for inhomogeneously distributed tft-lcd processes. Int. J. Prod. Res. **51**(18), 5644–5657 (2013)
24. Lin, W.-J., Chen, J.J.: Class-imbalanced classifiers for high-dimensional data. Briefings Bioinf. **14**(1), 13–26 (2013)
25. Lines, J., Bagnall, A., Caiger-Smith, P., Anderson, S.: Classification of household devices by electricity usage profiles. In: Yin, H., Wang, W., Rayward-Smith, V. (eds.) IDEAL 2011. LNCS, vol. 6936, pp. 403–412. Springer, Heidelberg (2011)
26. Liu, X.-Y., Zhou, Z.-H.: Ensemble Methods for Class Imbalance Learning. Wiley, New Jersey (2013)
27. López, V., Fernández, A., García, S., Palade, V., Herrera, F.: An insight into classification with imbalanced data: empirical results and current trends on using data intrinsic characteristics. Inf. Sci. **250**, 113–141 (2013)

28. MacDougall, P., Roossien, B., Warmer, C., Kok, K.: Quantifying flexibility for smart grid services. In: Power and Energy Society General Meeting (PES), 2013 IEEE, pp. 1–5, July 2013
29. Molina, J.M., Garcia, J., Garcia, A.C.B., Melo, R., Correia, L.: Segmentation and classification of time-series: real case studies. In: Corchado, E., Yin, H. (eds.) IDEAL 2009. LNCS, vol. 5788, pp. 743–750. Springer, Heidelberg (2009)
30. Pedregosa, F., Varoquaux, G., Gramfort, A., Michel, V., Thirion, B., Grisel, O., Blondel, M., Prettenhofer, P., Weiss, R., Dubourg, V., Vanderplas, J., Passos, A., Cournapeau, D., Brucher, M., Perrot, M., Duchesnay, E.: Scikit-learn: machine learning in python. J. Mach. Learn. Res. **12**, 2825–2830 (2011)
31. Piao, Y., Park, H.W., Jin, C.H., Ryu, K.H.: Ensemble method for classification of high-dimensional data. In: 2014 International Conference on Big Data and Smart Computing (BIGCOMP), pp. 245–249, January 2014
32. Roossien, B.: Mathematical quantification of near realtime flexibility for smart grids, flexines d8.1. Fproject Report, Energy research Centre of the Netherlands (ECN) (2012). http://www.flexines.org/publicaties/eindrapport/BIJLAGE14a.pdf
33. Saeys, Y., Inza, I.N., Larrañaga, P.: A review of feature selection techniques in bioinformatics. Bioinformatics **23**(19), 2507–2517 (2007)
34. Seo, M., Oh, S.: A novel divide-and-merge classification for high dimensional datasets. Comput. Biol. Chem. **42**, 23–34 (2013)
35. Sutton, C., Sindelar, M., McCallum, A.: Feature bagging: preventing weight under-training in structured discriminative learning. IR 402, Department of Computer Science, University of Massachusetts Amherst (2005)
36. Tax, D.: Ddtools, the data description toolbox for matlab, version 2.1.1, July 2014
37. Wang, D., Parkinson, S., Miao, W., Jia, H., Crawford, C., Djilali, N.: Hierarchical market integration of responsive loads as spinning reserve. Appl. Energy **104**, 229–238 (2013)
38. Wang, Z., Zhao, Z., Weng, S., Zhang, C.: Solving one-class problem with outlier examples by SVM. Neurocomputing, **149**, Part A: 100–105 (2015). Advances in neural networks Advances in Extreme Learning MachinesSelected papers from the Tenth International Symposium on Neural Networks (ISNN 2013) Selected articles from the International Symposium on Extreme Learning Machines (ELM 2013)
39. Wille-Haussmann, B., Erge, T., Wittwer, C.: Decentralised optimisation of cogeneration in virtual power plants. Sol. Energy **84**(4), 604–611 (2010). International Conference CISBAT 2007
40. Yang, H., King, I., Lyu, M.: Multi-task learning for one-class classification. In: The 2010 International Joint Conference on Neural Networks (IJCNN), pp. 1–8, July 2010
41. Yu, H., Mu, C., Sun, C., Yang, W., Yang, X., Zuo, X.: Support vector machine-based optimized decision threshold adjustment strategy for classifying imbalanced data. Knowl.-Based Syst. **76**, 67–78 (2015)

Correlation Analysis for Determining the Potential of Home Energy Management Systems in Germany

Aline Kirsten Vidal de Oliveira[1]([⊠]) and Christian Kandler[2]

[1] CAPES Foundation, Ministry of Education of Brazil, Brasilia, DF 70.040-020, Brazil
alinekvo@gmail.com
[2] Lehrstuhl für Energiewirtschaft und Anwendungstechnik, Technische Universität München,
Arcisstr. 21, 80333 Munich, Germany
christian.kandler@tum.de

Abstract. This paper describes the implementation of a model in MATLAB that estimates the potential of home energy management systems based on different component criteria. This is done by the estimation, in a given territory, of the correlation of favorable elements for the installation of integrated systems for energy generation and electromobility. The model is applied to the territories of Germany, evaluating its potential for home energy management systems in current and future situations.

Keywords: Electromobility · Renewable energy · Energy management systems

1 Introduction

Due to the current global concerns about CO2 emissions and dependency on fossil fuels, electric cars are seen as an alternative to traditional vehicles. However, they increase the load on the power grid, creating a need for changes in the power system, which is a major source of greenhouse gas. Thus, electric cars are only able to reduce emissions when renewable energy sources are used to charge them [1].

For this reason, the research project "Energy-autarkic electric mobility in the Smart-Micro-Grid" (Energieautarke Elektromobilität im Smart-Micro-Grid – e-MOBILie), aims to link electric mobility with local renewable power generation in an integrated approach. It will demonstrate, through simulations and a prototype, how to integrate electric vehicles into the energy management structure of intelligent buildings [2].

Research companies like Navigant Research, ABI Research and Grand View Research have made estimations of the potential of HEM systems (Home Energy Management Systems), by evaluating its revenue forecast. This market is likely to expand in the coming years as new products are introduced, especially in Europe where the growth will be more prominent, particularly due to British government incentives [3]. However, this kind of data does not estimate the installation potential of complete HEM systems and a method to do this is shown in this paper, based on the model developed in [4].

© Springer International Publishing Switzerland 2015
W.L. Woon et al. (Eds.): DARE 2015, LNAI 9518, pp. 94–104, 2015.
DOI: 10.1007/978-3-319-27430-0_7

The installation of a HEM system is facilitated in one-family households that already have one or more favorable elements, like a photovoltaic system (PV system), a garage with an electric car or a heat pump. Therefore, finding the number of houses that have these favorable elements means finding the potential for the installation of these systems. In order to do this, data on general characteristics of Germany related to the main project is used as input of the model, including predictions for the next thirty-five years. The work uses data about the amount and distribution of private photovoltaic systems, electric cars, houses with garages and private heat pumps.

The MATLAB model uses the number of one-family households from a territory as its domain where the elements are randomly distributed, what allows the visualization of possible correlations between them. By repeating this process several times, one can generate an estimation, for example, of the number of houses that have both existing photovoltaic systems and garages. Such houses have the potential to install a system for integration between in-house-generation and an electric car. The model described in this paper identifies these correlations and determines the stochastic distribution of the results. These results enable an analysis of the potential application of HEM systems in a particular region, thus indicating the applicability of the main project within a territory.

2 Data Search

For relevant results, the search for data input is one of the most important parts of the model development. The number of one-family households, private PV systems, private heat pumps, private electric cars and houses with garage are used as input for the model and they provide more accuracy when they are distributed in states. The input data used for the model covers the current situation of Germany and its forecasts for 2020, 2030 and 2050.

Some government organizations, like "Agentur für Erneuerbare Energie" and "Bayerische Staatsregierung" have interactive maps that present information about the distribution of renewable energies in Germany. However, some kinds of necessary data, like the number of existing houses with garages, cannot be found by the distribution in the area. For this reason, they are estimated using simple statistics or absolute values obtained in the literature.

The data are more accurate for the periods of 2015, 2020 and 2030, since there are many reports and statistics available that provide this kind of information. On the other hand, there are not so many reports about forecasts for 2050, and some extrapolations had to be done. Consequently, the data for 2050 is not so precise, especially because it ignores the lifetime of the equipment.

3 Model Description

The model developed in MATLAB correlates data to find the number of houses that have two or more elements favorable to the installation of HEM systems in a locality.

The model utilizes, as basis, the number of one-family-households in a given region divided by areas that can be states, cities or countries and uses it as the territory where the elements are randomly distributed and the overlays are collected. In this paper, three elements are explored, but the model is easily expandable for the use of more or less elements. In addition, it can be applied to any territory where the number of houses is available, from a neighborhood to the entire world. Figures 1, 2, 3 and 4 describe the steps of how the correlation process works.

After the steps are executed, the overlays are collected, which means that the model will register the number of times that a site has a combination of the distributed factors. Since the data is randomized, the process is executed several times, generating different correlations in each model run.

Fig. 1. Step 1: The model uses as domain the number of one-family households of each state. The figure shows an example with Germany as territory, where the draw of houses represents the number of one-family-households of each state.

Fig. 2. Step 2: The data from one element, like PV systems for example, is distributed in the territory. As it is not possible to know exactly each house has a PV system installed, this distribution is randomly made. The houses with PV Systems are presented in red (Color figure online).

Fig. 3. Step 3: Other element is distributed, for example electric cars, here in yellow. When the data of the element is distributed by areas (here states), the number of units is randomized in each area. Otherwise, the total number is randomized in the total territory. It may happen that some houses have a PV system and an electric car. These houses are the ones with potential for HEM systems and they are represented in orange (Color figure online).

Fig. 4. Step 4: When other element is distributed, for example heat pumps (blue), it is possible to collect the number of all the houses that have potential for HEM systems. They are houses with an electric car and a heat pump (green), an electric car and a PV system (orange), a heat pump and a PV system (purple) and an electric car, a PV system and a heat pump (grey) (Color figure online).

The sum of each run of the model is presented through histograms and a statistical analysis of the correlated data is performed, in order to determine its stochastic distribution. That means, finding a mathematical function that describes the statistical variable in an accurate way. This is largely used for describing phenomena in areas such as psychology, image processing and computer network traffic [5–7].

The software MATLAB has tools that fit the data to various stochastic distributions and give the statistics measures like Bayesian Information Criterion (BIC). This criterion estimates the accuracy of a fitting distribution, where the one with the lower BIC is the best fit. This model compares the result with the following distributions: beta, Birnbaum-Saunders, exponential, extreme value, gamma, generalized extreme value, inverse Gaussian, logistic, log-logistic, lognormal, Nakagami, normal, Rayleigh, Rician and t-location-scale. When the difference between the BIC of the best fit and the normal distribution is lower than six, the normal distribution is assumed as the best fit, since it is better-interpreted and only differences above six show strong evidence of a fit [8].

After a proper stochastic distribution is found, the distribution parameters are estimated based on the sample data. In this way, the curve can be easily reproduced.

Figure 5 demonstrates the result of the model for a German best-case scenario in 2020. The figure contains information about the data used and exhibits the histogram resulted from the model. In the histogram, it is possible to see that from the 100 iterations, in more than 20, the number of correlations between PV systems, heat pumps and electric cars were circa 3450, what means that a result of around 3450 is the most probable. In red, it is visible the approximation of the histogram to a normal distribution, which has its parameters listed. The figure displays also the criterions of approximation to the fit.

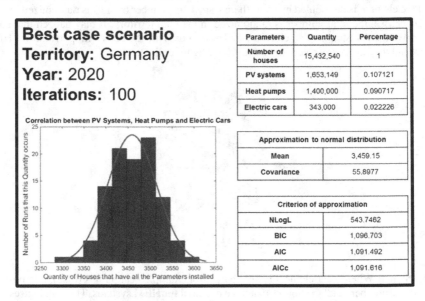

Fig. 5. Result of the model in a German best-case scenario in 2020 (Color figure online)

3.1 Dependencies

The model deals with total random data, but in reality, there are some correlations that should be considered. A person that already has a PV system installed at home, for example, is probably concerned about the environment and therefore is more susceptible to buy an electric car than others are. Because of that, some "dependencies" were added to the model, so according to the choice of the user, it is possible to increase the probability that a determined parameter can appear in a house that already has another one.

Figure 6 shows the results of a simulation with the same data input of the simulation of Fig. 5, but adding a dependency of 25 % from electric cars to PV systems. That means that the owners of private PV systems are 25 % more likely to buy electric cars. Therefore, the probability of them to buy an electric car is:

$$Probability\ of\ a\ PV\ system\ owner\ to\ buy\ an\ electric\ car = (1 + 0,25) *$$
$$Probability\ of\ anyone\ buy\ an\ electric\ car \tag{1}$$

The dependency causes differences in the results that can be seen in the histogram and in the mean of the normal distribution, which have raised. It is also perceptible that the mean of the normal distribution of the Fig. 5 is 3,459.15, result that is very close to the multiplication of the number of one-family households by the percentages of the elements, that is 3,333.18. With that alone, the model would be irrelevant, since similar results can be easily calculated. However, with the possibility of adding the dependencies between elements, the model proves its importance.

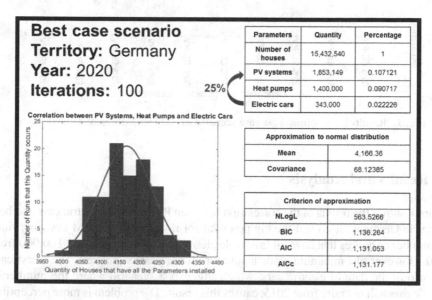

Fig. 6. Results for a German best-case scenario in 2020 with a dependency of 25 % from electric cars to PV systems

Since it is hard to find this kind of difference of probability, a sensibility analysis is done in order to obtain a range of the possible results. The following example, in Fig. 7, shows the results of the correlations between PV systems, heat pumps and electric cars for a German best-case scenario in 2020. In this scenario, it is assumed that one is more likely to have an electric car if they already have a heat pump installed. This increase in probability varies from 0 to 90 %, as showed in the figure bellow. This range of dependencies is simulated for each case testing different dependencies and a range of possible results is obtained.

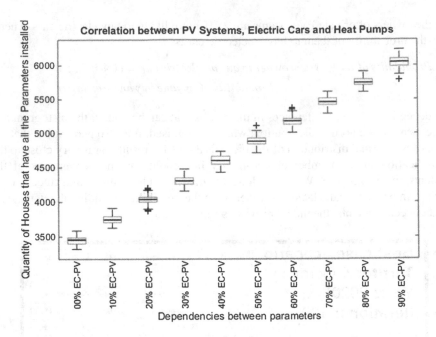

Fig. 7. Results for a German best-case scenario in 2020 with different dependencies

4 Results and Analysis

Figure 8 shows the results of the correlation between PV systems, electric cars and heat pumps in Germany. It is clear the big potential for installation of HEM systems, since the number of houses that have all favorable elements can be higher than 100,000 from 2030. However, the minimum value is low for all the years simulated. The worst-case scenario of the data of electric cars, which suggests a scenario where the number of electric cars will not raise from 2015, causes this result. The problem is more perceptible in the years of 2030 and 2050, when the difference between the worst-case and the other scenarios is bigger. It is possible to conclude that the correlation results are only probable with an increase in the number of electric cars.

From the model is also possible to obtain correlations combining only two elements, as seen in Figs. 9 and 10. They show combinations that are of interest of the project e-MOBILie, which links electromobily with local renewable generation, for the years of 2015 and 2020. The results present a big quantity of houses with a PV system and an electric car, which are great opportunities for the installation of HEM systems. Even today, there are at least 249 of these houses in Germany, where the first systems could be installed. In five years these number can reach 25,000 houses.

Table 1 presents all the ranges of the results obtained with the simulations. Here, it is clear the pessimistic worst-case scenarios obtained, as the differences between the biggest and the smallest results are very large.

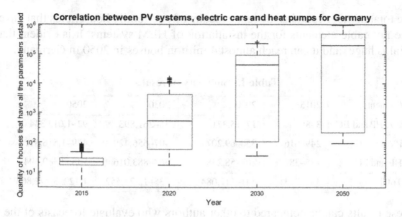

Fig. 8. Results of the correlation between PV systems, electric cars and heat pumps in Germany

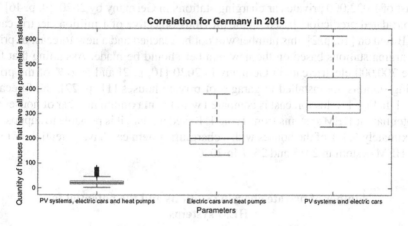

Fig. 9. Results of correlations for Germany in 2015

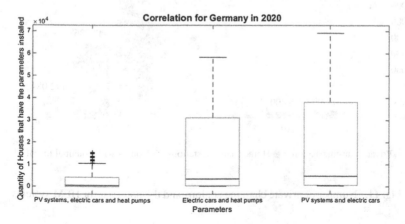

Fig. 10. Results of correlations for Germany in 2020

The total number contains the range results of one-family households that have two or three favorable elements for the installation of HEM systems. It is evident that the potential is huge, and it can reach almost 4 million houses in 2050 in Germany.

Table 1. Summary of results

Elements	2015	2020	2030	2050
PV, HP and EC	3–85	17–15,371	35–261,893	92–1,013,344
PV and EC	249–616	350–69,207	407–656,429	479–1,516,312
HP and EC	135–382	208–58,248	363–883,016	735–3,369,894
Total	381–913	541–112,084	735–1,277,552	1,122–3,872,862

These results can be compared to other authors who evaluate forecasts of the area. The report from the National Electromobility Platform (NPE), for example, shows a forecast of 1,022,000 private car charging stations in Germany by 2020. [9, p. 46] This is an old-dated prediction, based on the previous objective of 1 million electric cars in 2020. Based on [10, p. 2], this number will not be reached and a new forecast of private car charging stations based on the new number should be made. Assuming that there will be 700,000 electric cars in Germany in 2020 [10, p. 2] and 57.6 % of the private charging stations are installed in garages of private houses [11, p. 22], the forecast is updated. In Fig. 11, this forecast is compared with the maximum number of houses with the potential for HEM systems from Table 1. Based on that, it is possible to suppose that approximately 6.3 % of the houses with a charging system can have potential for installing a HEM system in 2015 and 25 % in 2020.

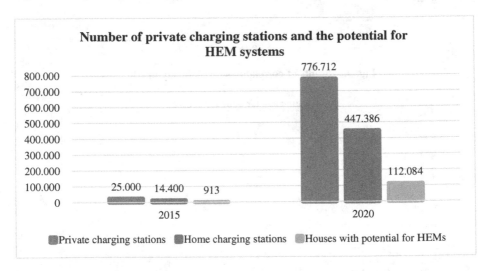

Fig. 11. Number of private charging stations and the potential for HEM systems

5   Conclusions

The model developed met the established objectives, being a reliable tool to estimate the potential of HEM systems, mainly when accurate data is available. A big amount of relevant data on factors related to the local generation of renewable energy and electromobility were gathered, what already illustrates the current and future scenario of this market. These data were correlated and analyzed, giving a good visualization of the possible application of the project e-MOBILie, also allowing the analysis of the future scenario. The model gives also the possibility of entering dependencies between elements and using ranges as inputs. Those things are the most interesting characteristics of the model as they differ it from estimations made with simple calculations.

It is possible to conclude that the market grown of HEM systems depends mostly from the success of the electric cars penetration. Since it succeeds, there is a big potential for the installation of these systems, with the possibility that the number of one-family households with at least two favorable elements to HEM systems installation reaches almost 4 million in 2050.

As proposed, the model is easily expandable and can be applied to other areas of study, as for example, the estimation of houses that have more than one technological devices (cellphones, laptops…). The expansion is also possible by raising the number of elements used in this research, adding factors such as wind turbines or battery banks installed. The model is also made to run to any territory, so this estimation can be expanded for other domains and even an evaluation of the world HEMS potential is possible. In order to do this, improvements in the velocity of the simulation shall be made.

Other further development could be to conduct surveys to have more accurate data for the inputs. To find out how much one is likely to buy an electric vehicle when they already have a PV system installed at home, for example, would turn possible to have one exact value for dependencies and to restrict the range of results. One other possibility is to use data that takes the life-time of the equipment into account or use statistics to estimate it. That would bring results that are more complete, especially for long-term forecasts like 2050. In addition, new trends, statistics and reports are realized periodically and the results obtained in this work can be frequently updated so estimations that are more precise are achieved.

References

1. Saber, A.Y., Venayagamoorthy, G.K.: Plug-in vehicles and renewable energy sources for cost and emission reductions. IEEE Trans. Ind. Electron. **58**, 1229–1238 (2011)
2. Lehrstuhl für Energiewirtschaft und Anwendungstechnik, "e-MOBILie - Schaufenster Elektromobilität," Lehrstuhl für Energiewirtschaft und Anwendungstechnik. http://www.ewk.ei.tum.de/forschung/projekte/e-mobilie/. Accessed 21 June 2015
3. Hill, J.S.: "Home Energy Management Revenue Set To Hit $22 Billion," CleanTechnica, 3 March 2015. http://cleantechnica.com/2015/03/03/home-energy-management-revenue-set-hit-22-billion/. Accessed 13 Mai 2015

4. Kirsten Vidal de Oliveira, A., Kandler, C.: Correlation Analysis for Determining the Potential of Home Energy Management Systems in Bavaria, Germany and Europe, Munich (2015)
5. Tayamachi, T., Wang, J.: Transmission characteristic analysis for UWB body area communications. In: International Symposium on Electromagnetic Compatibility (2007)
6. Li, B., Zhao, Z.: Cloth 3D attributes animation based on a wind noise. In: Symposium on Photonics and Optoelectronics (2012)
7. Cisar, P., Cisar, S.M.: Fitting univariate distributions to computer network traffic data using GUI. In: IEEE 13th International Symposium on Computational Intelligence and Informatics (CINTI) (2012)
8. Kass, R.E., Raftery, A.E.: Bayes factors. J. Am. Stat. Assoc. **90**, 773–795 (1995)
9. Nationale Plattform Elektromobilität (NPE), "Fortschrittsbericht 2014 – Bilanz der Marktvorbereitung," Gemeinsame Geschäftsstelle Elektromobilität der Bundesregierung (GGEMO), Berlin (2014)
10. Hacker, F., von Waldenfels, R., Mottschall, M.: "Wirtschaft lichkeit von Elektromobilität," February 2015. http://ikt-em.de/_media/Gesamtbericht_Wirtschaftlichkeit_von_Elektromobilitaet.pdf. Accessed 27 March 2015
11. National Organisation Hydrogen and Fuel Cell Technology "Ergebnisbericht - Der Modellregionen Elektromobilität 2009–2011," Bundesministerium für Verkehr, Bau und Stadtentwicklung, Berlin (2012)

Predicting Hourly Energy Consumption. Can Regression Modeling Improve on an Autoregressive Baseline?

Pierre Dagnely[1]([✉]), Tom Ruette[1], Tom Tourwé[1], Elena Tsiporkova[1], and Clara Verhelst[2]

[1] Sirris - Software Engineering and ICT Group, Brussels, Belgium
{Pierre.Dagnely,Tom.Ruette,Tom.Tourwe,Elena.Tsiporkova}@sirris.be
[2] 3E - iLab, Brussels, Belgium
Clara.Verhelst@3e.eu

Abstract. According to the Third Industrial Revolution, peer-to-peer electricity exchange combined with optimized local storage is the future of our electricity landscape, creating the so-called "smart grid". Such a grid not only has to rely on predicting electricity production, but also its consumption. A growing body of literature exists on the topic of energy consumption and demand forecasting. Many contributions consist of presenting a methodology, and showing its accuracy. This paper goes beyond this common practice on two levels: first, by comparing two regression techniques to a univariate autoregressive baseline and second, by evaluating the models in term of industrial applicability, in close collaboration with domain experts. It appears that the computationally costly regression models fail to significantly beat the baseline.

Keywords: Energy consumption prediction · Ordinary least squares regression · Support vector regression · Autoregressive models

1 Introduction

According to the Third Industrial Revolution defined by Rifkin [26], peer-to-peer electricity exchange combined with optimized local storage is the future of our electricity landscape. Therefore, a "smart grid" not only has to rely on predicting electricity production, but also consumption. Moreover, the prediction of hourly electricity consumption is of great economical interest for players on the global electricity market, since an accurate prediction of the consumption is needed to obtain the best prices on the day-ahead market and to avoid purchasing on the more expensive real-time spot market. This study[1] focuses on the prediction of the (hourly) electricity consumption of a medium-scale office building. Through its Software-as-a-Service platform *SynaptiQ*, 3E offers an interface

---

[1] This study takes place in the context of the Artemis project Arrowhead (http://www.arrowhead.eu), in which 3E and Sirris collaborate.

© Springer International Publishing Switzerland 2015
W.L. Woon et al. (Eds.): DARE 2015, LNAI 9518, pp. 105–122, 2015.
DOI: 10.1007/978-3-319-27430-0_8

to monitor (green and grey) electricity production and consumption, and is now additionally interested in predicting electricity consumption.

Energy consumption prediction is a typical forecasting challenge, where a univariate continuous variable (hourly total energy consumption in a building) needs to be predicted by means of a model that may contain multiple exogenous variables, i.e. independent variables that affect a model without being affected by it, and endogenous variables, i.e. variables that affect a model but are also affected by it. For the current paper, regression models are investigated as candidate methods for the forecasting challenge. Concretely, the research question is formulated as "which regression model is most successful in predicting as precisely as possible the total energy consumption in a building". Thus, two regression techniques, Ordinary Least Squares and Support Vector Regression, are compared against a naive autoregressive baseline. It appears that it is (for the given realistic dataset) impossible to significantly improve on that autoregressive baseline with the two established regression models, despite their increased computational cost.

The paper first offers a literature review of energy consumption forecasting work, and some references for a better understanding of regression models (Sect. 2). Then, the data and the considered predictors are discussed in detail (Sect. 3). In the following section, the baseline against which the results of the regression models will be compared and the two regression models (Sect. 4) are presented. Then, an evaluation of these models against a held-out test dataset and an investigation of the improvement over the baseline is presented (Sect. 5). The penultimate section discusses these results (Sect. 6). Finally, Sect. 7 concludes the paper and offers a list of topics for further research.

2   Literature Review

Predicting the electricity consumption of buildings has been a topic in the literature for at least fifty years [31], during which mainly three approaches have been explored: the engineering approach, the statistical approach, and the machine learning approach.

Engineering methods use physical models that are based on building characteristics, such as the wall materials or the HVAC[2] characteristics of each room, and external parameters, such as the weather, to predict electricity consumption. These methods range from simple manual estimations to detailed computational simulations [4].

An example of a simple engineering method is the bin or temperature frequency method. This method starts by defining outdoor temperature bins of usually 2 or 3°C. For each bin, the electricity consumption of the building is estimated, assuming that the consumption depends on the outdoor temperature. By multiplying the estimated electricity consumption for each temperature bin with the amount of times that this frequency bin occurs, the total electricity

---

[2] HVAC stands for Heating, Ventilation and Air-Conditioning.

consumption is estimated. To have a more accurate prediction, there can be multiple electricity consumption estimates for a temperature bin, e.g. an estimate for electricity consumption during business hours and an estimate outside of the business hours [4].

On the other side of the spectrum, a detailed computational simulation would estimate the electricity consumption based on a finely grained picture of all the building components and characteristics by means of an existing tool. A list of such tools is maintained by the US department of energy [1]. These tools usually follow four steps [4]:

– Estimate the cooling and heating consumption for each zone of the building.
– Estimate the required energy flows of the various equipments to obtain this cooling and heating consumption.
– Determine the amount of electricity that is consumed to generate these energy flows.
– Estimate the cost of this electricity consumption (optional).

Two strategies exists to tackle these steps:

– Sequential strategy: all steps are handled one after the other, using the result of the previous step as input.
– Integrated strategy: all steps are performed together, with feedback loops between them. This strategy is more complex but takes into account the interactions between the steps.

An example of a computational simulation tool is *EnergyPlus*, an integrated simulation system sponsored by the US Department of Energy and first released in 2001. EnergyPlus is composed of three basic components [10]. The first component is the simulation manager, which manages all the processes. The second component is the heat and mass balance simulation module, which computes the building thermal zones and their interactions. This module can use parameters such as room surface, heat conduction of walls (based on their sizes and materials) or daylight illumination. The third component is the building system simulation module, which simulates the HVAC and electrical systems, equipments and components to update the zone-air condition and estimate the electricity consumption.

Statistical approaches are based on methods coming from the classical statistician toolbox. Typically, historical data that represents the past behaviour of a building, e.g. past consumption, weather or sensor data such as occupancy, is used to fit a model, e.g a regression model. However, predicting electricity consumption can be a time series problem, where data points at time t depend on data points at a time $t - i$. Therefore, methods such as Auto Regressive Integrated Moving Average (ARIMA) are widely used, as to account for the interdependency of data points. These ARIMA methods consist basically of two models. First, an autoregressive model assumes that a value at time t is linearly depending on values at times $t - i$ and a noise term. Second, a moving average model assumes a stationary mean, and future values in the time series are functions of this mean, altered by weighted noise terms. In [24], Newsham et al. evaluated an ARIMAX model, i.e. an ARIMA model with eXogenous inputs that influence the noise terms, for predicting the electricity consumption of a three-story building of $5800\,m^2$, comprising laboratories and 81 individual work space. Due to the presence of laboratories, this building does not represent a conventional office building. As predictors, they used past consumption data, weather data and occupancy data (through sensor monitoring logins but not logoffs). Adding the login predictor improved slightly the accuracy of the model. Nevertheless, they conclude that measuring logoffs or better sensors, such as a camera, may have a positive impact on the accuracy in a more conventional office building.

Regular regression analyses have also shown good accuracy, despite the potential violation of the assumption of independence among data points. In [6], Ansari et al. computed the cooling consumption of a building by means of a linear regression function with the temperature difference between inside and outside as a predictor. Cho et al. studied the impact of the training dataset size when applying linear regression to predict the energy consumption of a building in [9]. They used the average outdoor temperature as a predictor for the energy consumption of the heating system. With one day of hourly measurement data, they obtained a very inaccurate system with an error range from 8 to 117 %. A training dataset of one week of daily data gave an error range of 3 to 32 %, while for three weeks of training data the errors ranged from 9 to 26 %. They concluded that in a training set shorter than one month the outdoor temperature variability is a more important cause of error than the length of the training set. In contrast, with a training set of more than one month of daily data, the length of the measurement period strongly influences the accuracy. As an example, with three months of daily data, their model was able to make electricity consumption predictions with errors ranging from only 1 to 6 %.

Machine learning approaches are, like statistical approaches, data-driven, but they use techniques coming from the field of Artificial Intelligence. Two of the most used methods in energy consumption prediction are Artificial Neural Networks [28] and Support Vector Machines [11]. Artificial Neural Networks create a network of input nodes, in-between nodes, and output nodes, all connected by weighted links. The output nodes are thus a function of the input nodes, but their relationship can be obfuscated by hidden in-between nodes, and the relationships created by the links. Support Vector Machines are typically used for classification problems. To resolve non-linear classification problems, Support Vector Machines transpose the data points to a higher dimensional space by means of a kernel. In this higher dimensional space, the data points may have a linear separation. The linear separation is then found by fitting a hyperplane between the classes. The hyperplane has to maximize the margins, i.e. the largest distance to the nearest training data point from the classes. Support Vector Regression [28] is an extension of a Support Vector Machine adapted to regression problems. Here, the hyperplane has to minimize the error cost of the data points outside of the margin ε, while the error cost of the data points inside the margin are considered as null. It prioritize the minimisation of the upper bound of the training error rather than the (global) training error.

In [13], Ekonomou used Artificial Neural Network to predict the (yearly) Greek long-term energy consumption (2005–2015) using the past thirteen years as training data (from 1992 to 2004). The inputs for fitting the model were the yearly ambient temperature, the installed power capacity, the yearly per resident electricity consumption and the gross domestic production. He tested various models, using the combination of 5 back-propagation learning algorithms, 5 transfer functions and 1 to 5 hidden layers with 2 to 100 nodes in each hidden layer. The best model had a compact structure and a fast training procedure with 2 hidden layers of 20 and 17 nodes, using the Levenberg-Marquardt back-propagation learning algorithm and logarithmic sigmoid transfer function. Gonzalez et al. [14] used an Auto-associative feed-forward Neural Network to predict the electricity consumption of two buildings, with input variables such as the temperature difference $\triangle T = T_{k+1} - T_k$, the hour of the day, the day of the week and the previous consumption. Gonzalez et al. showed that the previous consumption reflects other parameters such as the occupancy level: their models performed relatively accurately during recurring holidays, although this information was not directly encoded as predictor.

For the past ten years, Support Vector Machines are on the rise and many studies have been performed. One of the earliest applications of a Support Vector Machine to the (monthly) forecast of building energy consumption is [12], where Dong et al. successfully applied it to four commercial buildings with a better accuracy than other machine learning methods. They emphasized on the fact that Support Vector Machines only require the tuning of a few parameters: ε, C and the kernel parameters (such as γ for a Radial Basis Function kernel). In a more recent study, Jain et al. [17] examined the impact of various spatial and temporal variables to predict the energy consumption of a multi-family residential building. They applied a Support Vector Machine with the following predictors: the past consumption, temperature, solar flux, weekend/holidays and hour of the day. They tested the impact of spatial (i.e. by apartment, by floor or for the whole building) and temporal (i.e. every 10 min, hourly or daily) granularity of the data. The model that was based on data per floor and per hour was found to be significantly more accurate than the other possible combinations.

Over the years, some studies that compare these three types of approaches have been performed. In [23], Neto et al. compared the aforementioned *EnergyPlus* tool to a machine learning approach based on Feed-forward neural networks. The result of this comparison revealed that both approaches are similar in term of prediction accuracy. However, the Feed-forward neural network approach turned out to be more straightforward, as it only relied on 17 months of past consumption and weather data. In contrast, the engineering model was cumbersome to construct, as it depended on the availability of domain experts and a precise knowledge of all the building characteristics.

Azadeh et al. [7] compared Artificial Neural network models using a supervised multi layer perceptron network to a conventional regression technique for predicting the monthly electricity consumption in Iran. The more accurate Artificial Neural Network model outperformed the conventional regression significantly. For the specific case of ARIMA models, studies shown that they are usually easy to estimate, but lack in accuracy in comparison with machine learning methods [31]. However, Amjady [5] tested a modified ARIMA taking into account the domain knowledge of experts (e.g. to define manually a starting point for the parameter tuning). Artificial Neural Network and ARIMA where similar in term of accuracy but the modified ARIMA succeeded to outperform slightly both models.

From a mathematical perspective, and in terms of optimization, Support Vector Machines present some advantage in comparison with Artificial Neural Network [12]. Support Vector Machine methods always lead to a unique and globally optimal solution, whereas (back propagation) Artificial Neural Network methods may lead to a locally optimal solution. From an application perspective, both Artificial Neural Networks and Support Vector Machines produce accurate results, but need a sufficient amount of representative historical data to train the model. Usually, however, Support Vector Machine models are slightly more accurate, as shown in [2].

Many other machine learning approaches have been compared. Tso et al. [29] applied decision trees, which have the advantage of being easily interpretable (in contrast to Artificial Neural Networks and Support Vector Machines). Huo et al. [16] experimented with genetic programming, which mimic the natural evolution to make evolutionary programs by recombination, mutation and selection to find good solutions. Yang et al. [30] studied the use of fuzzy rules, through a combination of Wang-Mendel method and a modified Particle Swarm Optimization algorithm.

However, a promising approach could be the use of a hybrid method, i.e. a combination of different methods. Hybrid methods may be able to go beyond the accuracy of a Support Vector Machine or Artificial Neural Network, as shown in [18]. In that study, Kaytez et al. combined Support Vector Machine with Least Squares as a loss function to construct the optimization problem. Nie et al. [25] investigated the combination of ARIMA, to predict the linear part of the consumption and Support Vector Machine to predict its non-linear part. Li et al. [20] tested a General Regression Neural Network, i.e. an Artificial Neural Network where the hidden layer consists of Radial Basis Functions.

From this extensive body of literature, three aspects are missing, which we try to address in the current paper. First, there has been, up to now, no comparison of ordinary least squares regression versus support vector regression. Second, none of the presented studies compare the more advanced techniques to the simplest possible baseline. Third, all methods have only been evaluated in terms of accuracy, but not in terms of industrial applicability, i.e. complexity to deploy and maintenance of the system. This last evaluation step has been performed in close collaboration with the domain experts of 3E.

3   Response and Predictor Variables

For this study, a 72 hours-ahead prediction of the hourly electricity consumption of the 3E office in Brussels was requested by 3E. They provided five years of past electricity consumption data (from 2010 to 2014). In addition, this study was performed in close collaboration with experts at 3E, who are not only experienced with energy consumption metrics, but also know the context and activity of the 3E office. This proved to be very useful to identify certain predictive patterns in the data that emerged during explorative visualizations.

This chapter starts by an explorative analysis of the received data. Then the response variable, i.e. the variable to predict, is explained more deeply. Finally, the predictors tested during this study are described in detail, with a focus on how they have been implemented.

The following building specific patterns have been found:

- A drop in energy consumption was visible between 13:00 and 14:00, which, after consultation with a 3E employee, appeared to align with the typical lunch break moment in the building. Since the effect was relatively small, the lunch break was not modelled as a predictive variable.
- Between 2010 and 2013, energy consumption on Saturdays was higher than on Sundays, and in 2014 this pattern was not visible. It appeared that between 2010 and 2013, the cleaning company worked in the building on Saturdays, and in 2014 this policy changed. Because of this very clear policy change, 2014 has been left out from this analysis.
- In 2013, a new electrical heating system was installed, which increased the electricity consumption to a certain extent. Since this effect was relatively small, 2013 has not been discarded from the analysis.

3.1   Response Variable

The response variable will be the hourly energy consumption. The histogram in Fig. 1 shows the distribution of the hourly energy consumption measurements at a granularity of 100 W, which is the measurement granularity.

3.2   Predictors

Some potentially relevant predictors have been selected, based on the analysis of the data and a literature review on the predictors. In the following subsections, these predictors and how they have been incorporated in the benchmark data set, during the preprocessing step, is discussed.

Recency. An autocorrelation analysis of the electricity consumption shows a clear weekly pattern. In addition, an autocorrelation with the previous day and all the $t - (\alpha \times 7)$ previous days with a decreasing impact over times has also been observed. Therefore, three autoregressive attributes have been created:

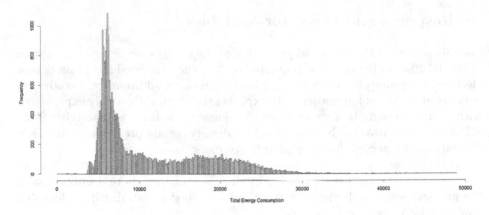

Fig. 1. Histogram of total energy consumption at 100 W granularity.

RECENCY1 that merely contains the logged energy consumption value of the previous day, RECENCY7 that contains the logged energy consumption value of exactly one week ago and RECENCY14 that contains the value of exactly two weeks ago. These recency attributes also correspond to the autoregressive baselines used to evaluate the added value of the more advanced regression models.

Temporal Predictors. As defined in [19], the first temporal attribute considered is the typical OCCUPANCY of the building. For the 3E office building, standard business hours range from 08:00 until 19:00. It is obvious that in this period, the total energy consumption will be higher than outside of this period. In addition, during weekend days, the facilities are typically not used. This information has been encoded as a binary attribute. In principle, the OCCUPANCY attribute may also contain planned holidays, or even foreseeable late night events, or general off-site meetings. However, this additional information has not been taken into account since it was not readily available.

The second temporal attribute is the day of the week. Visual inspection has revealed that there are recurring patterns of reduced/increased energy consumption that relate to the specific day of the week. This can be explained by habits of employees, who have typical days for teleworking. This information is encoded as the categorical WEEKDAY variable that contains simply the name of the day of the week. The encoding of this information is based on the visual inspection of data, but can also be based on the outcome of a model that can be inspected, such as a decision tree.

Meteorological Predictors. As shown in [8], two meteorological attributes can be sufficient to integrate the weather into a predictive model. The first meteorological attribute is (ambient) TEMPERATURE. The outside temperature has an influence on total energy consumption, since on cold days, additional electrical heating may be used. On warm days, the air conditioning might be responsible for increased energy consumption. 3E provided the hourly measured ambient temperature in Brussels (Uccle), where their office building is located. The TEMPERATURE has been encoded as a continuous variable.

The second meteorological attribute is IRRADIANCE. The irradiance, which is the amount of sunlight that reaches the earth, may influence total energy consumption. A lower irradiance may indicate darkness or cloudiness, which increases the need for artificial lighting. Also, there is an influence on the heating. 3E provided the hourly measured irradiation in Brussels (Uccle). The IRRADIANCE has also been encoded as a continuous variable. As expected, there is a strong correlation to the temperature predictor.

4 Modeling

As mentioned in the introduction to this paper, two regression approaches (Ordinary Least Squares and Support Vector Regression) have been compared to a naive autoregressive baseline. It is expected that the Ordinary Least Squares regression and Support Vector Regression models will predict the hourly electricity consumption more accurately than an autoregressive baseline, because they take the temporal and meteorological factors into account, in addition to the recency attributes.

The data set has been divided between a test dataset and a training dataset. The training dataset covers the two years 2011 and 2012. The year 2010 has been discarded, because there was no meteorological data available. The test dataset, which is never used during the training phase, consists of data from the year 2013.

4.1 Autoregressive Baseline

To evaluate the quality of the regression models, three autoregressive models have been evaluated and the most accurate one has been used as a baseline. These three models predict the energy consumption at time t by taking the energy consumption at respectively times $t - 1$ day, $t - 7$ days and $t - 14$ days, because these are the three strongest autocorrelation lags. Note that these components have also been used as the recency attributes in the regression models.

4.2 Regression

Based on a review of the literature, two regression methods have been selected. The first regression method considered is Ordinary Least Squares regression [15], a statistical method. The optimal linear combination of predictors is found by fitting a hyperplane between the data points. This hyperplane has to minimize the sum of the squares of the distances from the points to this hyperplane. This regression method is straightforward to implement, free of parameters and can be run without making any configuration decisions.

The second regression method considered is Support Vector Regression [28], a machine learning method. Unfortunately, an exact method to obtain the optimal parameters of a Support Vector Machine does not exist. Therefore, a search algorithm must be applied. Three types of search algorithm approaches exist:

1. Grid search, where a set of possible parameter values is tested, i.e. for each parameter, a range of possible values is assessed and all combinations are tested. Such an approach is used by Akay in [3]. This approach produces good result, but it usually is a time consuming method.
2. Local search type methods, such as the pattern search applied by Momma et al. [22], where locally optimal Support Vector Machine models are created and then bagged or averaged to produce the final model.
3. A more recent approach uses machine learning methods to estimate the parameters. For example, Salcedo-Sanz et al. [27] compared Evolutionary Programming and Particle Swarm Optimization to find the parameters of a Support Vector Regression model for a wind speed forecasting problem. Both methods had very good performance, but such methods are more complex to deploy than a simple grid search.

For this study, the grid search approach has been used. Since the Radial Basis Function kernel has been chosen — it was successfully applied by Dong et al. [12] — three parameters have to be tuned (on the training set): C and ε, which are Support Vector Machine related parameters, and γ, which is a kernel related parameter. For the model using all predictors, the grid search suggested as parameters: kernel = Radial Basis Function (RBF), $C = 100000, \gamma = 0.03$ and $\varepsilon = 0.00005$.

For both methods, 7 different combinations of predictors have been generated to find out which interactions yield the most accurate results. The first four models only take a single predictor, respectively OCCUPANCY, TEMPERATURE, IRRADIANCE and RECENCY7. The fifth model considers the two temporal attributes and the two meteorological variables in interaction (OCCUPANCY, WEEKDAY, TEMPERATURE and IRRADIANCE). The sixth model uses the three recency attributes in interaction (RECENCY 1, RECENCY 7, RECENCY 14). The seventh model considers all attributes (RECENCY 1, RECENCY 7, RECENCY 14, OCCUPANCY, WEEKDAY, TEMPERATURE and IRRADIANCE) in interaction.

5   Evaluation

Figure 2 illustrates nicely how well the predicted values estimate the observed values, and how close to one another the predictions of the baseline, Ordinary Least Squares regression and Support Vector Regression are. The figure depicts a seven days-ahead forecast based on one year of training data and using all the predictors aforementioned for the Ordinary Least Squares regression and Support Vector Regression models.

The predictive power of the models have been evaluated by letting them forecast the hourly (total) energy consumption for the test dataset. For each day of

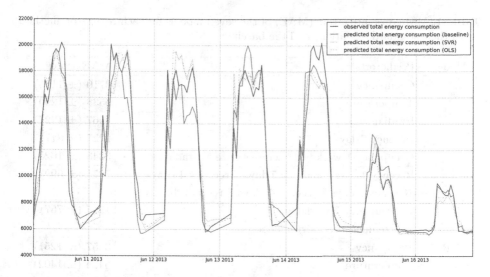

Fig. 2. Showcase of prediction performance.

the test dataset a forecast of the three next days have been made (72-hours ahead prediction is requested by 3E) and the prediction accuracy of this forecast has been calculated by using the Mean Absolute Error (MAE). Note that these predictions rely on actual meteorological data, instead of relying on predicted temperature and irradiance values, so that the results tend to be over-optimistic. Then, the mean and standard deviation of all these 72-hour ahead prediction accuracies across the whole year have been computed. The MAEs of the 7 models from both the Ordinary Least Squares regression and the Support Vector Regression are presented in Table 1. The same scores have been computed for the three autoregressive models. The model using the energy consumption at $t - 7$ days appeared as the more accurate and has been chosen as a baseline.

The MAEs only provide a global view on the errors in the prediction. To analyze the errors, three checks have been undertaken (1) a check of the distribution of the errors, (2) an analysis of errors to identify common patterns that can be addressed in the model, (3) a comparison of the MAE distribution for the main models.

First, the distribution of the (absolute) errors of the prediction have been inspected – based on the Support Vector Regression model with all predictors (the errors of the Ordinary Least Squares regression prediction are similarly distributed) – by means of a histogram, represented in Fig. 3. Taking into account that the errors are similarly distributed in the baseline, the Ordinary Least Squares regression model and the Support Vector Regression model, the MAE is a decent metric for comparison.

Second, the significant deviations between observed and predicted values have been explored to identify recurring patterns, by computing the MAE scores of the main models for various specific time periods, as shown in Table 2. For

Table 1. Mean absolute errors (MAE) of the seven models for each regression method (with their standard deviation) and the baseline

Model	Predictor(s)	MAE scores
OLS	Occupancy	3710 (± 1195)
	Temperature	5538 (± 1290)
	Irradiance	4967 (± 1227)
	Recency 7 days	2227 (± 817)
	Occupancy * weekday * temperature * irradiance	3343 (± 1025)
	Recency 1 days * recency 7 days * recency 14 days	1971 (± 699)
	Recency 1 days * recency 7 days * recency 14 days * occupancy * weekday * temperature * irradiance	1914 (± 757)
SVR	Occupancy	3657 (± 1251)
	Temperature	5424 (± 1402)
	Irradiance	4630 (± 1561)
	Recency 7 days	2125 (± 837)
	Occupancy * weekday * temperature * irradiance	3219 (± 1071)
	Recency 1 days * recency 7 days * recency 14 days	1914 (± 691)
	Recency 1 days * recency 7 days * recency 14 days * occupancy * weekday * temperature * irradiance	1719 (± 596)
Baseline	Recency 1 days	3242 (± 1287)
	Recency 7 days	2189 (± 871)
	Recency 14 days	2391 (± 1016)

both regression flavors, it can be observed that the prediction errors appear (to be expected) mainly during the working hours, and not during the night. A similar distinction can be observed between the workweek and the weekend in the errors. However, as the electricity consumption of the night and the weekend is significantly lower, the percentage of the error is higher for these periods, e.g. a Support Vector Regression model using all predictors has an error of 16.6 % for the weekend and 10.4 % for the workweek. Further inspection points to a slight increase in errors during the spring and the winter periods for both regression models whereas the prediction error of the autoregressive baseline is similar for all seasons. This difference of accuracy in spring can be attributed to increased energy consumption during more dynamic, extreme warm or cold, weather. For winter, 5 weeks of data are missing, which could impact the accuracy of the models and lead to this result.

Third, a comparison of the distribution of MAE values for the main models have been conducted, as shown in Fig. 4. It can be observed that models only based on meteorological and temporal predictors are less accurate than

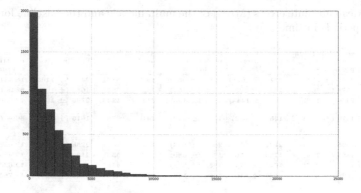

Fig. 3. Histogram of prediction errors of SVR.

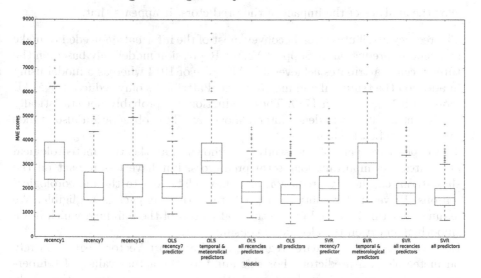

Fig. 4. Distribution of MAE values for the main models

the others. There is no significant difference (in the sense that confidence intervals overlap) between the Ordinary Least Squares regression, Support Vector Regression models and the baseline.

For the calculation of the models, the following Python libraries have been used: Statsmodels for the Ordinary Least Squares regression, SciKit-Learn for the Support Vector Regression and Pandas for the data manipulation.

6   Discussion

The evaluation of the models (see above) points out that the following two aspects have an impact on the prediction results: (1) the selection of the predictors for regression models, (2) the selection of a modeling method.

Table 2. Mean absolute errors (MAE) of the main models and the baseline for forecast of specific periods in time

	OLS all recencies predictors	OLS all predictors	SVR all recencies predictors	SVR all predictors	Recency 7 baseline	Mean of the electricity consumption
Workweek	2145 (± 951)	2112 (± 1028)	2090 (± 960)	1884 (± 877)	2415 (± 1165)	18.031 (± 5403)
Weekend	1578 (± 765)	1480 (± 693)	1488 (± 754)	1390 (± 781)	1680 (± 1095)	8343 (± 3225)
Working hours	2476 (± 1362)	2354 (± 1444)	2407 (± 1363)	2140 (± 1243)	2799 (± 1672)	14.938 (± 6504)
Night and weekend	1570 (± 753)	1459 (± 612)	1535 (± 749)	1410 (± 678)	1616 (± 903)	8094 (± 3295)
Spring	2095 (± 726)	2122 (± 959)	2092 (± 813)	1911 (± 786)	2258 (± 1036)	12.618 (± 5877)
Summer	1888 (± 794)	1739 (± 656)	1779 (± 680)	1572 (± 536)	2190 (± 958)	13.478 (± 7310)
Autumn	1829 (± 627)	1645 (± 474)	1758 (± 573)	1590 (± 475)	2071 (± 713)	12.634 (± 5992)
Winter	2030 (± 506)	2147 (± 610)	1998 (± 514)	1885 (± 519)	2166 (± 520)	13.766 (± 6732)

For the analysis of the impact of the predictors, it appears that:

- The recency attributes already convey most of the information needed to make an accurate prediction. A Support Vector Regression model only based on the three recency attributes achieves a MAE score of 1914 whereas a model using in addition the temporal and meteorological attributes only reduces the MAE score with 195, to reach 1719. These reduction are probably not statistically significant as their confidence intervals overlap. This observation also stands for Ordinary Least Square.
- The explorative part of the study has shown that the two meteorological attributes are important long-term predictors, but have less impact on the short-term, e.g. hourly prediction and they relate well to the meteorological seasons. However, for short-term prediction, their impact is negligible. We assume that this is caused by the thermal inertia of the building, which has a smoothing effect on the electricity consumption.
- Both Ordinary Least Squares regression and Support Vector Regression rely on meteorological predictors. For this analysis, the actual values of temperature and irradiance have been used. However, in a real application, only predictions of these values will be available. This might decrease slightly the efficiency of both regression methods (but has no influence on the baseline).

Based on the above findings, we would like to argue that a regression model can be successfully applied, even if the previous consumption is the only available data. This result is particularly interesting for companies working at the grid level as the recency data is the only one that they directly have access to. They could therefore do without purchasing expensive weather forecasts or to study the occupancy of the buildings monitored.

For the comparison of regression models with an autoregressive method, it appears that:

- The naive baseline already produces good results, with a mean absolute error of 2189. Given a mean hourly energy consumption of 13099 W at the 3E building, this is an error of 16.7 %.

- Support Vector Regression and Ordinary Least Square Regression manage to improve slightly on this baseline. They respectively reach a mean absolute error of 1719 (13,1 %) and 1914 (14,6 %). However, the confidence intervals for these mean absolute error scores overlap, which suggests that improvement will not be statistically significant.
- Both Ordinary Least Squares regression and Support Vector Regression are slow in constructing the model on the basis of the training data. The baseline does not need this intensive computation step, and will return predictions much faster.

These points are also particularly interesting from an industrial point of view as they imply that a very simple autoregressive method already gives good results. In addition to the obvious simplicity of this method in comparison to more complex regression models, this method has the advantage to be easily deployable and maintainable by people without specific data science expertise, as it only relies on a simple preprocessing of the data.

7 Conclusion

In this study a comparison of Ordinary Least Square and Support Vector Machine models with an autoregressive method have been performed to make a 72 hours-ahead forecast of the electricity consumption of an office building. We conclude that both the autoregressive baseline as the more advanced regression models are able to predict the hourly energy consumption fairly well. However, the advanced regression models do not significantly outperform the autoregressive baseline. Given the computational cost of the advanced regression models, an autoregressive model is the most effective methodology at present to do energy consumption prediction. The temporal and meteorological predictors do not improve the accuracy significantly in our tests.

7.1 Further Research

This study is a reflection of our first explorations of the prediction of hourly energy consumption. Next, we want to further explore the application of Support Vector Regression by testing other kernels, by estimating the parameters more dynamically, i.e. recalibrating the model when concept drift occurs, or by using preprocessing methods, e.g. binning (a first investigation did not show much improvement) or wavelets.

In addition, we want to explore more advanced autoregressive models, using a weighted average of past values. The intuition behind the weighted average is that the recency lags do not have the same importance. It can be expected that the electricity consumption of 7 days ago is closer to the future consumption than the one of 14 days ago. The consumption of one day ago has also less importance than might be assumed at first: the presence of daily patterns tends to minimize it. As an example, Friday is usually a day where employees tend to telework, which decreases the electricity consumption. Therefore, basing the forecast on the consumption of Thursday could lead to an overestimation. The problem is more obvious for the prediction of the electricity consumption of Mondays based on the consumption of Sundays. An initial inspection with a weighted average of the three recency attributes has already produced encouraging results, with a MAE score of 1984 ($\pm$ 740). This MAE score is close to the one of the Ordinary Least Square model only using recency attributes as their weights are relatively similar and as the interaction between the predictors do not have a big impact in the Ordinary Least Square model.

We also want to test if other methods could significantly out-perform the autoregressive models. One such method is k Nearest Neighbour, a method as simple as the autoregressive method. To predict the electricity consumption of day_{i+1}, this method takes the $days_{\{i-j,i\}} = H$, find k sequences of j days that resemble H (the so-called k nearest neighbours, i.e. the k sequences of days with the closest electricity consumption behaviour to H that are present in the historical data) and then take, for each sequence, the subsequent day, this day would correspond to a potential prediction for day_{i+1}. The forecast is made with a weighted average of these following day of the k nearest neighbours, using weights based on the similarity of these neighbours with H. This method was successfully applied by Lora et al. [21], who used it to make a 24 hours-ahead forecast of the spanish electricity demand.

Acknowledgements. This work was subsidised by the Region of Bruxelles-Capitale - Innoviris.

References

1. Building Technologies Office: Building Energy Software Tools Directory
2. Ahmad, A.S., Hassan, M.Y., Abdullah, M.P., Rahman, H.A., Hussin, F., Abdullah, H., Saidur, R.: A review on applications of ANN and SVM for building electrical energy consumption forecasting. Renew. Sustain. Energy Rev. **33**, 102–109 (2014)
3. Akay, M.F.: Support vector machines combined with feature selection for breast cancer diagnosis. Expert Syst. Appl. **36**(2), 3240–3247 (2009)
4. Al-Homoud, M.S.: Computer-aided building energy analysis techniques. Build. Environ. **36**(4), 421–433 (2001)
5. Amjady, N.: Short-term hourly load forecasting using time-series modeling with peak load estimation capability. IEEE Trans. Power Syst. **16**(3), 498–505 (2001)
6. Ansari, F.A., Mokhtar, A.S., Abbas, K.A., Adam, N.M.: A simple approach for building cooling load estimation. Am. J. Environ. Sci. **1**(3), 209–212 (2005)
7. Azadeh, A., Ghaderi, S.F., Sohrabkhani, S.: Forecasting electrical consumption by integration of neural network, time series and ANOVA. Appl. Math. Comput. **186**(2), 1753–1761 (2007)
8. Beccali, M., Cellura, M., Lo Brano, V., Marvuglia, A.: Short-term prediction of household electricity consumption: assessing weather sensitivity in a Mediterranean area. Renew. Sustain. Energy Rev. **12**(8), 2040–2065 (2008)
9. Cho, S.-H., Kim, W.-T., Tae, C.-S., Zaheeruddin, M.: Effect of length of measurement period on accuracy of predicted annual heating energy consumption of buildings. Energy Convers. Manage. **45**(18–19), 2867–2878 (2004)

10. Crawley, D.B., Lawrie, L.K., Pedersen, C.O., Winkelmann, F.C., Witte, M.J., Strand, R.K., Liesen, R.J., Buhl, W.F., Joe Huang, Y., Henninger, R.H., et al.: EnergyPlus: new, capable, and linked. J. Architectural Plann. Res. 21(4), 292–302 (2004). Theme Issue: Advances in Computational Building Simulation (Winter, 2004)

11. Cristianini, N., Shawe-Taylor, J.: An Introduction to Support Vector Machines and Other Kernel-based Learning Methods. Cambridge University Press, New York (2000)

12. Dong, B., Cao, C., Lee, S.E.: Applying support vector machines to predict building energy consumption in tropical region. Energy Build. 37(5), 545–553 (2005)

13. Ekonomou, L.: Greek long-term energy consumption prediction using artificial neural networks. Energy 35(2), 512–517 (2010)

14. González, P.A., Zamarreño, J.M.: Prediction of hourly energy consumption in buildings based on a feedback artificial neural network. Energy Build. 37(6), 595–601 (2005)

15. Hamilton, J.D.: Time Series Analysis, vol. 2. Princeton University Press, Princeton (1994)

16. Huo, L., Fan, X., Xie, Y., Yin, J.: Short-term load forecasting based on the method of genetic programming. In: International Conference on Mechatronics and Automation, ICMA 2007, pp. 839–843. IEEE (2007)

17. Jain, R.K., Smith, K.M., Culligan, P.J., Taylor, J.E.: Forecasting energy consumption of multi-family residential buildings using support vector regression: Investigating the impact of temporal and spatial monitoring granularity on performance accuracy. Appl. Energy 123, 168–178 (2014)

18. Kaytez, F., Cengiz Taplamacioglu, M., Cam, E., Hardalac, F.: Forecasting electricity consumption: a comparison of regression analysis, neural networks and least squares support vector machines. Int. J. Electr. Power Energy Syst. 67, 431–438 (2015)

19. Kwok, S.S.K., Lee, E.W.M.: A study of the importance of occupancy to building cooling load in prediction by intelligent approach. Energy Convers. Manage. 52(7), 2555–2564 (2011)

20. Li, Q., Ren, P., Meng, Q.: Prediction model of annual energy consumption of residential buildings. In: 2010 International Conference on Advances in Energy Engineering (ICAEE), pp. 223–226. IEEE (2010)

21. Lora, A.T., Santos, J.M.R., Riquelme, J.C., Expósito, A.G., Ramos, J.L.M.: Time-series prediction: application to the short-term electric energy demand. In: Conejo, R., Urretavizcaya, M., Pérez-de-la-Cruz, J.-L. (eds.) CAEPIA/TTIA 2003. LNCS (LNAI), vol. 3040, pp. 577–586. Springer, Heidelberg (2004)

22. Momma, M., Bennett, K.P.: A pattern search method for model selection of support vector regression. In: SDM, pp. 261–274. SIAM (2002)

23. Neto, A.H., Fiorelli, F.A.S.: Comparison between detailed model simulation and artificial neural network for forecasting building energy consumption. Energy Build. 40(12), 2169–2176 (2008)

24. Newsham, G.R., Birt, B.J.: Building-level occupancy data to improve ARIMA-based electricity use forecasts. In: Proceedings of the 2nd ACM Workshop on Embedded Sensing Systems for Energy-Efficiency in Building, pp. 13–18. ACM (2010)

25. Nie, H., Liu, G., Liu, X., Wang, Y.: Hybrid of ARIMA and SVMs for short-term load forecasting. Energy Procedia 16, 1455–1460 (2012)

26. Rifkin, J.: The third industrial revolution: How the internet, green electricity, and 3-d printing are ushering in a sustainable era of distributed capitalism. World Financial Review, 1 (2012)
27. Salcedo-Sanz, S., Ortiz-Garcı'a, E.G., Pérez-Bellido, Á.M., Portilla-Figueras, A., Prieto, L.: Short term wind speed prediction based on evolutionary support vector regression algorithms. Expert Syst. Appl. **38**(4), 4052–4057 (2011)
28. Smola, A.J., Schölkopf, B.: A tutorial on support vector regression. Stat. Comput. **14**(3), 199–222 (2004)
29. Tso, G.K.F., Yau, K.K.W.: Predicting electricity energy consumption: a comparison of regression analysis, decision tree and neural networks. Energy **32**(9), 1761–1768 (2007)
30. Yang, X., Yuan, J., Yuan, J., Mao, H.: An improved WM method based on PSO for electric load forecasting. Expert Syst. Appl. **37**(12), 8036–8041 (2010)
31. Zhao, H., Magoulès, F.: A review on the prediction of building energy consumption. Renew. Sustain. Energy Rev. **16**(6), 3586–3592 (2012)

An OPTICS Clustering-Based Anomalous Data Filtering Algorithm for Condition Monitoring of Power Equipment

Qiang Zhang[1(✉)], Xuwen Wang[2], and Xiaojie Wang[3]

[1] State Grid Electric Power Research Institute, Beijing 100192, China
zhangqiang7@sgepri.sgcc.com.cn
[2] Institute of Medical Information, CAMS & PUMC, Beijing 100020, China
wang.xuwen@imicams.ac.cn
[3] Beijing University of Posts and Telecommunications, Beijing 100876, China
xjwang@bupt.edu.cn

Abstract. In allusion to the widespread anomalous data in substation primary equipment condition monitoring, this paper proposes an OPTICS (Ordering Points To Identify the Clustering Structure) clustering-based condition monitoring anomalous data filtering algorithm. Through the characteristic analysis of historical primary equipment condition monitoring data, an anomalous data filtering mechanism was built based on density clustering. The effectiveness of detecting anomalous data was verified through the experiments on one 110 kV substation equipment transformer oil chromatography and the GIS (Gas Insulated Substation) SF6 density micro water. Compared with traditional anomalous data detection algorithms, the OPTICS Clustering-based algorithm has shown significant performance in identifying the features of anomalous data as well as filtering condition monitoring anomalous data. Noises were reduced effectively and the overall reliability of condition monitoring data was also improved.

Keywords: Anomalous data · OPTICS clustering · Condition monitoring · Data mining

1 Introduction

In recent years, along with the development of electric power equipment intelligent communication, the acquisition of highly reliable data of substation primary equipment is required in smart grid. However, the anomalous data of equipment condition monitoring may affect the reliability of data.

At present, the main causes of anomalous data in condition monitoring included: (1) the breakdown of signal acquisition part occurs frequently, such as failure of sensor. (2) the high failure rate of communication in the substation. (3) the poor anti-interference ability of measurement system and (4) the failure of transmission or processing of data. The large number of anomalous data may bring negative effect to the condition monitoring system, fault diagnosis system and the reliability of data analysis. Therefore, filtering the anomalous data effectively becomes an important research problem in the smart grid.

© Springer International Publishing Switzerland 2015
W.L. Woon et al. (Eds.): DARE 2015, LNAI 9518, pp. 123–134, 2015.
DOI: 10.1007/978-3-319-27430-0_9

The traditional methods are mainly based on threshold [1] and the statistic of measurement of outliers, such as three sigma criteria (Pauta Rule). These principles were based on the scope of a certain accuracy to detect outliers [2]. However, the outlier data distribution characteristic and regularities in electric power equipment field was different from statistical methods in the distribution of the default assumption. In the real application, these methods can filter out some part of the normal data or even real failure data of equipment, which lead to the accuracy degradation of acquisition data and reduce the performance of fault diagnosis.

In this paper, the clustering technique was first applied in the analysis of anomalous data of condition monitoring, we proposed an OPTICS clustering-based anomalous data filtering algorithm [3]. First, we utilized OPTICS density clustering algorithm for mining the distribution characteristics of acquisition data. Then, we designed an anomalous data filtering strategy, combining with electric power equipment state standards and threshold decision rules, etc. Compared with traditional methods, our method excavated potential characteristics distribution of condition monitoring data of substation equipment, and filtered more anomalous data effectively.

2   Related Work

Data processing of power transmission and transformation equipment condition monitoring was divided into monitoring data upload, data storage, and data alarm analysis [4]. Anomalous data was processed before data storage, as shown in Fig. 1:

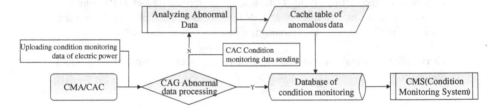

Fig. 1. The work flow of processing electric equipment condition monitoring anomalous data

Power equipment monitoring devices in substation, which access to the CAG (Condition Acquisition Gateway) using CMA (Condition monitoring agent) and CAC (Condition Acquisition Controller) devices, achieve the condition information of equipment. The main station system integrated condition monitoring data and performed the secondary processing of monitoring data, including the anomalous data filtering, threshold calculation of monitoring data [5] and data storage.

2.1   Threshold Method

The traditional method of condition monitoring data acquisition uploaded data to data servers directly through the IEC61850 communication protocols without the data preprocessing step. However, data processing was increasingly important.

Recently, some preprocessing schemes showed effectiveness in reducing noises, such as the Pauta Rule [6], see as formula (1). $\bar{x}$ was the sample mean, δ was the sample variance.

$$|x_d - \bar{x}| > 3\delta. \tag{1}$$

2.2 Statistic Method

Statistical methods for the anomalous data detection, such as the Grubbs test [7], the Dixon test [8] and the t–test, often assumed that the sample data fit the independent normal distribution, and filtered anomalous data according to the level of significance test and confidence interval. The main difference between them was the test statistic distribution and the critical value table.

Taking the Grubbs test formula as an example for anomalous data detection, see as follows:

$$G_{(n)} = \left(X_{(n)} - \bar{x}\right)/s \tag{2}$$

$$G'_{(n)} = \left(\bar{x} - X_{(1)}\right)/s \tag{3}$$

$G_{(n)}$: function that $x_{(i)} > \bar{x}$. $G'_{(n)}$: function that $x_{(i)} < \bar{x}$. $\bar{x}$: Sample mean. S: The standard deviation.

3 Our Method

3.1 Anomalous Data

Electric power equipment data was not a scalar. It was necessary to evaluate the outliers of a data object collection, which had multiple dimension parameters. There were multiple causes of anomalous data of the device object, such as the sensor fault, communication failure, collecting device failure and so on.

As shown in Fig. 2, taking the equipment operating data of 2# substation oil chromatography part parameters as an example, the red areas of real-time data represented a typical anomalous condition, since all of their parameters had a zero value. The yellow areas in Fig. 2(a) showed another type of anomalous condition, which has abnormal C_2H_4 parameter (zero value) and normal CH_4 value. It was difficult to judge the current condition of the equipment by one or two parameters.

Observing the occurrence of anomalous data, which had high frequency at the initial operating stage, we made a definition among different parameters for anomalous data of power equipment condition monitoring:

Definition 1: Electric power equipment condition anomalous data was the collection of observed object, which have significant differences or inconsistent performance with the whole data set of the equipment [9].

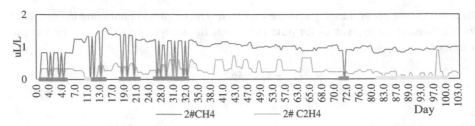

(a) Anomalous data distribution of 2# transformer oil chromatography methane and ethylene parameter

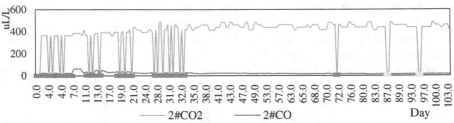

(b) Anomalous data distribution of 2# transformer carbon monoxide, carbon dioxide parameter

Fig. 2. The historical sample data of a substation transformer oil chromatography equipment condition, the red and yellow areas were two kinds of anomalous data.

Definition 2: Electric power data outlier degree referred to the degree of deviation between the object data and the whole or local data set.

3.2 An OPTICS Clustering-Based Anomalous Data Filtering Algorithm

Unlike partitioning cluster, such as k-means algorithm, the OPTICS [12] algorithm was not sensitive to dirty data or anomalous data. It was also less dependent on the neighborhood radius of initial parameter and the min points of neighborhood. OPTICS algorithm generated an order of augmented cluster for clustering analysis. The order represented the density-based cluster structure of various samples. In other words, it can get DBSCAN clustering results based on the any parameter and MinPts from the generated sequence. Like DBSCAN, OPTICS requires two parameters: ε, which describes the maximum distance (radius) to consider, and MinPts, describing the number of points required to form a cluster. A point P is a core point if at least MinPts points are found within its ε-neighborhood N_ε (p). Contrary to DBSCAN, OPTICS also considers points that are part of a more densely packed cluster, so each point is assigned a core distance that describes the distance to the MinPts-th closest point:

$$core - dist_{\varepsilon,minPts}(p) = \begin{cases} \text{Undefined} & \text{if } |N_\varepsilon(p)| < MinPts \\ MinPts \text{ - th smallest distance to } N_\varepsilon(p) & \text{otherwise} \end{cases} \quad (4)$$

The reachability-distance of another point o from a point p is the distance between o and p, or the core distance of p:

$$reachablity - dist_{\varepsilon,minPts}(o,p) = \begin{cases} \text{Undefined} & \text{if } |N_\varepsilon(o)| < MinPts \\ \max(core \text{ - } dist_{\varepsilon,MinPts}(p), dist(p,o)), & \text{otherwise} \end{cases} \quad (5)$$

If p and o are nearest neighbors, this is the $\varepsilon' < \varepsilon$ we need to assume in order to have p and o belong to the same cluster. Both the core-distance and the reachability-distance are undefined if no sufficiently dense cluster (w.r.t. ε) is available. Given a sufficiently large ε, this will never happen, but then every ε-neighborhood query will return the entire database, resulting in $O(n^2)$ runtime. Hence, the ε parameter is required to cut off the density of clusters that is no longer considered to be interesting and to speed up the algorithm this way.

Pseudo code:

```
OPTICS(datasets, eps, MinPts)
    for each point p : datapoints
        p->reachability-distance = UNDEFINED
    for each unprocessed point p of datasets
        N = getNeighbors(p, eps)
        mark p as processed
        output p to the ordered queue
        if (core-distance(p, eps, Minpts) != UNDEFINED)
            Seeds = empty priority queue
            update(N, p, Seeds, eps, Minpts)
            for each next q in Seeds
                N' = getNeighbors(q, eps)
                mark q as processed
                output q to the ordered list
                if (core-distance(q, eps, Minpts) !=
UNDEFINED)
update(N', q, Seeds, eps, Minpts)
```

The flow chart of OPTICS clustering-based condition monitoring anomalous data filtering algorithm was shown in Fig. 3:

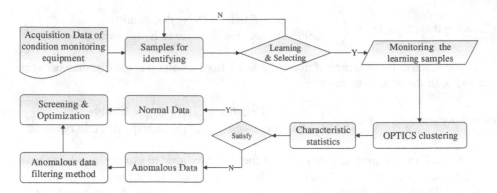

Fig. 3. Flow chart of condition monitoring anomalous data filtering algorithm

The main steps:

- Obtain condition acquisition data object as learning samples.
- Sort the data object by OPTICS clustering algorithm and output the clusters by density.
- Calculate the characteristic probability of the sorted clusters results: according to the OPTICS clustering results of ordered sample object id, calculate the number of core object clusters, the cluster distances and the similarity between clusters.
- According to the sample quantity within clusters and cluster similarity, sorting clusters from high to low order, it can distinguishes the anomalous cluster from other clusters.
- Filter out the anomalous clusters, optimize the learning samples.

4   Experiment

4.1   Transformer Oil Chromatographic Anomalous Data Analysis

Data. The Experiment chose nearly 3 years transformer 2# DGA (Dissolved Gas-in-oil Analysis) of an 110 kV step-down substation as sample data. The amount was nearly 4000. The anomalous data was generated in several periods. In the early operation, the cause of anomalous data was instability. The anomalous rate of data was higher and displayed irregular distribution. After that, the anomalous data displayed a periodic distribution.

The main measure standard of transformer condition monitoring parameters of anomalous data was as follows.

- Monitoring parameters value of the object were all zero;
- The whole or local parameter value of object data deviated from the mean value.
- Local monitoring parameters (such as CO CO_2) of object data were zero.

Based on the above definition of anomalous data, we can judge the anomalous condition of equipment.

Clustering Analysis. We obtained the object set of sorted clusters by OPTICS, setting $\varepsilon = 2$, MinPts $= 5$. Core objects were sorted based on density. Table 1 shows the sample of cluster results. Each row represents a core object vector. The second column represents the number of sorted sample objects in each cluster. Items in Column 3 to 10 are normalized values of parameters in the core object vector, such as CH_4, C_2H_6, C_2H_4, etc. It was verified by professional maintainers that the more objects the cluster contained, the more possible it was a normal cluster.

Table 1. The OPTICS cluster density core objects of transformer oil chromatographic sample

Cluster id	Quantity of objects	CH_4	C_2H_6	C_2H_4	C_2H_2	H_2	CO	CO_2	Total hydro-carbon
0	42	0.0068	0	0.00067	0	0.03007	0.11486	0	0.00753
1	358	0.0022	0	0.00062	0	0.01046	0.04952	0.93	0.00289

Verification of Algorithm

(1) *Recognition accuracy* P_a. The anomalous data accuracy was defined as follows:

$$P_a = \frac{N_{a,t}}{N_{x,t}} \tag{6}$$

$N_{a,t}$: The number of normal acquisition data under timing constants
$N_{x,t}$: All samples under the same timing constants.
The recognition accuracy P_a is a ratio between normal samples and the total data. As shown in Fig. 4, according to the sorting result of OPTICS clustering algorithm, we obtained the condition statistics table of equipment anomalous data distribution. Value "1" represented the normal condition of equipment and "0" represented the anomalous condition.

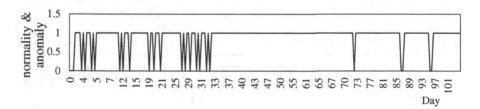

Fig. 4. The data result of OPTICS Clustering, 1: normal object 0: anomalous object

As shown in Fig. 5, the Pauta Rule and the Grubbs test gave the contrast curves of normal and anomalous data. The x axis represented sampling time series. The y axis represented each monitoring parameter anomalies. Each color curve in high position was in the normal condition. The number in y axis were used to separate each parameter in positions. The comparison of Fig. 5 (a) and (b) showed the in-conformity of anomalous distribution. For example, when the parameter CH_4 was anomalous, other parameters should also be anomalous. If the anomalous data object were filtered out based on partial attributes, some normal data may also be eliminated.

The transformer oil chromatography was given as an example to verify the OPTICS. Four methods were used to calculate the recognition accuracy: (1) our method; (2) traditional methods (including Pauta Rule, Grubbs and Dixon), as shown in Table 2. Our method improved the recognition accuracy rate by 30 %.

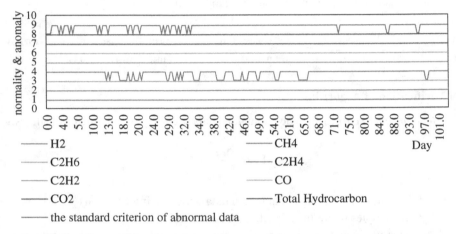

(a)   The status contrast chart of data in Pauta Rule

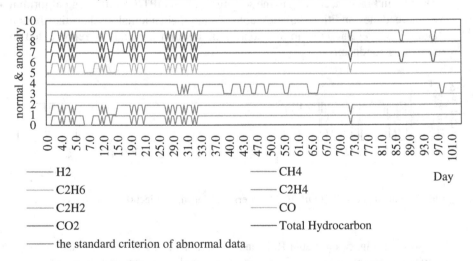

(b)   The status contrast chart of data in Grubbs test method

Fig. 5. Statistical comparison of data parameters in Pauta and Grubbs test method, $n = 3$ $\alpha = 0.01$ (Color figure online)

Table 2. The comparison of transformer oil chromatogram condition monitoring anomalous data recognition accuracy

Time	Pauta Rule	Grubbs	Dixon	Our method
4 months	0.65	0.5875	0.7208	**0.9833**
8 months	0.6577	0.5676	0.7336	**0.9712**
1 year	0.4315	0.5030	0.6278	**0.9818**

(2) *Internal measurement* [10]. The structure of power equipment condition monitoring data sets was unknown. The evaluation of clustering results depended on the characteristic and value of the data set. Under the situation, the measurement of anomalous data clustering analysis was referred to the tightness and separation degree. In addition, the size of a single cluster should also be considered. The minimization of inner-cluster distance was achieved by variance within the cluster. Based on the anomalous data filtering algorithm, the historical data curve of the parameters showed better smoothness and stability in Fig. 6.

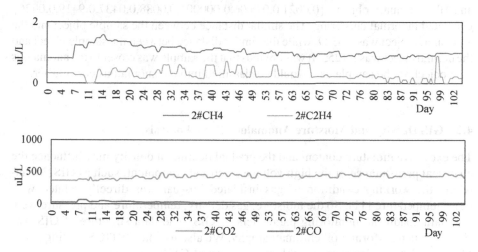

Fig. 6. Historical data curve of 2# substation transformer oil chromatogram monitoring parameters based on OPTICS anomalous data filtering algorithm

(3) *Special case of anomalous analysis.* A group of typical anomalous data objects were taken for further analysis of the OPTICS anomalous data filtering algorithm. Checking by the professional electric power equipment maintainers, the anomalous data points of one-year oil chromatogram condition were listed in Table 3.

Table 3. Typical anomalous samples of transformer oil chromatogram gas component condition monitoring data

Data object id	H$_2$ (uL/L)	CH$_4$ (uL/L)	C$_2$H$_6$ (uL/L)	C$_2$H$_4$ (uL/L)	C$_2$H$_2$ (uL/L)	CO (uL/L)	CO$_2$ (uL/L)	Total hydro-carbon (uL/L)
396	4.375	0.959	0	0.217	0	17.259	0	1.176
399	3.858	0.847	0	0	0	14.811	0	0.847
439	3.858	0.931	0	0	0	14.405	0	0.931
440	3.898	0.903	0	0.162	0	14.675	0	1.065

In theory, the value of H2, CO, CO$_2$ in Table 3, which were real data in operation period, should not be zero. However, the acquisition data value was zero. The traditional methods such as threshold value and the statistical magnitude failed to recognize the zero value as outliers. Our method used OPTICS clustering to sort and obtain the density of the certain data object: {0.0283, 0, 0.0061, 0, 0.1012, 0.5212, 0, 0.0344}.

We compared the sample data object with the anomalous object {0,0,0,0,0,0,0,0} and the normal object {0.0021,0,0.0006,0.000001,0.0088,0.0437,0.9419,0.0026} generated in formal clustering. The similar distance between the sample object and the anomalous object was 0.5329, while the similar distance between the sample object and the normal object was 1.1255. It was obvious that the sample was closer to the anomalous core object, so the sample data would be classified into the anomalous core cluster.

4.2 GIS Density and Moisture Anomalous Data Analysis

The excessive moisture content and the gradual decline of density may influence the electrical performance of the high voltage electrical equipment, such as GIS. Meanwhile, the working condition of gas insulated bus-bar was directly related with inner-temperature [11]. So the moisture, gas density, temperature and pressure were the main condition monitoring parameters of GIS. The data structure of GIS was similar with transformer oil chromatography. We also use the OPTICS, setting $\varepsilon = 2$, MinPts = 5, to analyze the anomalous condition of GIS equipment.

As shown in Tables 4 and 5, the anomalous data of moisture based on OPTICS algorithm, compared with the Pauta Rule, Grubbs test, and Dixon test, was different from transformer oil chromatogram results. The recognition accuracy decreased at first, and then increased. It was found that the temperature sensor of GIS SF6 contained anomalous data filtering mechanism. In first four months, the data collected from sensors were in the normal range. After 8 months, the growth of anomalous data was caused by communication failure.

Table 4. GIS SF6 density micro water anomaly data recognition accuracy rate

Time	Pauta rule	Grubbs	Dixon	Our method
4 months	1	0.519 2	0.798 0	0.663 4
8 months	0.915 5	0.8	0.853 3	0.844 4
1 year	0.997 0	0.714 2	0.746 3	0.831 3

Table 5. Samples of GIF SF6 moisture data normalized OPTICS core objects

Cluster id	Quantity of objects	Temperature	Density	Pressure under 20° C	Dew point	Moisture
0	69	0.068	0.082	0.013	−0.43	0.378
1	163	0.063	0.103	0.002	−0.6	0.227
2	168	0.045	0.125	0.002	−0.74	0.089

5 Conclusion

This paper proposed the OPTICS-based anomalous data filtering algorithm for condition monitoring of electric power equipment. Our method was compared with threshold method and statistical method in real-data experiments.

Experimental results showed that our method has effectively filtered out the anomalous data of power equipment condition monitoring. The average accuracy of anomalous recognition was 87 %.

Some anomalous data samples, which were too difficult to be identified by traditional methods, were caught by our method. So the proposed algorithm showed better effect than traditional methods. It was believed to be more suitable for multidimensional condition monitoring data collection.

In addition, the parameter ε in OPTICS was often set by experience. Normally, the sample data contained both normal data and anomalous data. However, if there was no anomalous data in the sample data, the OPTICS algorithm may produce an average distribution of the ordered density core objects.

In further work, we will keep on optimizing the abnormal data filtering strategy of the power condition monitoring equipment, and carry out further comparison experiments with more clustering algorithms.

References

1. Production and Technology Department of State Grid Q/GDW169-2008 Guide for condition evaluation of oil-immersed power transformers(reactors). China Electric Power Press, Beijing (2008) (in Chinese)

2. State Economic and Trade Commission DL/T722—2000 Guide to the analysis and the diagnosis of gases dissolved in transformer oil. China Electric Power Press, Beijing (2001) (in Chinese)
3. Ester, M., Kriegel, H.-P., Sander, J., Xu, X.: A density-based algorithm for discovering clusters in large spatial databases with noise. In: Simoudis, E., Han, J., Fayyad, U.M. (eds.). Proceedings of the Second International Conference on Knowledge Discovery and Data Mining (KDD-1996), pp. 226–231. AAAI Press (1996). ISBN 1-57735-004-9
4. Zhang, S.-Z., Yao, J.-Q.: Research on abnormal condition monitoring data filtering and alarm mechanism. Electr. Power Inf. Commun. Technol. (1) (2013)
5. Y-Q, C., K-P, L.: A condition monitoring method of generators based on RBF dynamic threshold model. Proc. CSEE **27**(26), 96–101 (2007)
6. Li, J.-L., Zhou, L.-K.: Detecting and identifying gross errors based on "3σ Rule". Comput. Modernization **1**, 10–13 (2012)
7. Osorio, F., Paula, G.A., Galea, M.: On estimation and influence diagnostics for the Grubbs' model under heavy-tailed distributions. J. Comput. Stat. Data Anal. **53**, 1249–1263 (2009)
8. McDaniel, W.C., Dixon, W.J., Walls, J.T.: Examination of Dixon's up and down method for small samples in the estimation of the ED50 for defibrillation. In: Engineering in Medicine and Biology Society, Proceedings of the Annual International Conference of the IEEE, vol. 13, pp. 760–761. IEEE (1991)
9. Zhang, X., Chen, M., Xiao F.: Origin used in comparison the methods of eliminating the excrescent data. Exp. Sci. Technol. (1), 74–76 (2012)
10. Zhang, W., Liu, C., Li, F.: Method of quality evaluation for clustering. Comput. Eng. **31**(20), 10–12 (2005)
11. Zhuang, J., Ke, M., Qin, W.: Research on SF6 gas density and moisture online monitoring for high-voltage apparatus. Computer Measurement & Control (2013)
12. Mihael, A., Markus, M.B., Hans-Peter, K., Jörg, S.: OPTICS: ordering points to identify the clustering structure. In: ACM SIGMOD International Conference on Management of Data, pp. 49–60. ACM Press (1999)

Argument Visualization and Narrative Approaches for Collaborative Spatial Decision Making and Knowledge Construction: A Case Study for an Offshore Wind Farm Project

Aamna Al-Shehhi, Zeyar Aung$^{(\boxtimes)}$, and Wei Lee Woon

Institute Center for Smart and Sustainable Systems (iSmart),
Masdar Institute of Science and Technology, Abu Dhabi, United Arab Emirates
{amalshehhi,zaung,wwoon}@masdar.ac.ae

Abstract. A Geographic Information Systems (GIS) play a vital role in various applications associated with sustainable development and clean energy. In these applications, the GIS provides a capability to upload on-site geographical information collected by public into online maps. One of the major problems is how to make a decision for those reports. In this paper, we study two types of cognitive modes for decision making: argument and narrative reasoning. We investigate how various discussion representations, argumentation theoretical model, and reasoning modes of geo-graphics problems affect knowledge accumulation and argument quality. We conduct empirical tests on different groups of participants regarding their discussions on a particular offshore wind farm project as a case study. We have demonstrated that graph representation provides better results than threaded representation for collaborative work. We also illustrate that the argument theoretical model leads to reduce participants' performance. Moreover, we conclude that there is no significant difference between narrative representation and graph representation in the participants' performance to construct knowledge.

Keywords: Argumentation models · IBIS · GIS · Narrative · story-telling · Public participation · PPGIS

1 Introduction

Geographic Information Systems (GIS) play a vital role in various applications associated with business efficiency, clean energy, sustainable development, disaster response and global climate change [10]. In these applications, GIS can provide a facility to upload on-site geographic information collected by crowd-sourcing to online maps. The online maps enable tracking of reports of events in different locations in order to help responders or interested parties provide resources, solutions and decision making for those events. Those events consider as complex or wicked problems because there is not a right or wrong solution for it. In particular, geo-graphics planning processes which involve many factors

© Springer International Publishing Switzerland 2015
W.L. Woon et al. (Eds.): DARE 2015, LNAI 9518, pp. 135–154, 2015.
DOI: 10.1007/978-3-319-27430-0_10

such as environmental, economic and social. Moreover, there are various unseen consequences of any proposed solutions [32]. When stakeholders collaborate to discuss the problem, they reveal a range of aspects of the problem depending on their own expertise, interest and background. This provides a range of different opinions and possible solutions [32]. Participant interaction can carry out during the planning process which starts with the problem and end with finding the best solution [38].

In our everyday life, we normally solve problems or make decisions by an "argumentation process". Argumentation can be defined as "verbal and social activity of reason aimed at increasing (or decreasing) the acceptability of a controversial standpoint for the listener or reader" [22]. In this sense, argument can carry the decision making process [29]. In the age of the Internet, the argumentation process has been digitized in different forms over the web such as emails, forums, chatting rooms and blogs. Alternatively to the argumentative reasoning, there is a narrative reasoning. The argument approach reaches one conclusion and the process is objective in nature. On the other hand, the narrative approach has more than one conclusion and it is subjective - based on the individual person's experience. Moreover, the narrative mode is used for exploring ambiguous issues to find a solution [2]. Numerous mapping tools have been developed during the last decade. Those tools involve public participation in the planning process by providing comments through a forum. However, making the information available does not ensure coming up with the best decision about a problem. Moreover, most of the current spatial decision-making tools are not integrated Web 2.0 technology [26]. As a result, the current tools are not suitable for deliberating around high controversy or complex problems such as energy and climate change [12]. This drive this study in which the current geo-graphic decision-making tools make the GIS component and data available, but having data available is not enough to come up with the best solution as indicate above. UN and FEMA emphasizes that the ability of presenting the information in a consumable way in the situation have a positive consequence [35]. In this work, we proposed the following research questions.

- **Research question 1**: How does the use of constrain-based argumentation scaffolds using IBIS model argument model affect users' online knowledge construction and argument quality about high controversy problems?
- **Research question 2**: How do different representations (visualization) affect collaborative users' knowledge construction and argument quality about high controversy problems?
- **Research question 3**: Does reasoning approaches affect the knowledge produce about high controversy problems? We will provide the answers to the proposed research questions by conducting empirical study.

In order to address the above research questions, we conducted a two-phase experiment with 37 participants by requesting them to discuss about the "Cape Wind Offshore Energy Project" in Nantucket Sound, Massachusetts, USA using different argumentation methods and later assessing the quality of discussions.

2   Background

This section presents a brief background on argumentation (Sect. 2.1). It also explores narrative and story-telling reasoning (Sect. 2.2). Finally, we present a brief introduction about Public Participant Geographic Information System (Sect. 2.3).

2.1   Argumentation

Argumentation is a "verbal and social activity of reason aimed at increasing (or decreasing) the acceptability of a controversial standpoint for the listener or reader" [22]. Argument is a human activity reflected in our daily lives, for example in talk shows, and during discussions and chats. With the spread of the Internet, this practice is digitalized in different forms such as emails, forums, chat rooms and blogs. Argumentation is used to resolve daily conflicts existing in our online and offline interaction [28]. Even though argumentation is a natural human activity, it is a skill that can be improved [37]. There are many components or ontologies that integrate to structure the argumentation. Each component has a role and function to serve. Argument in its general form consists of a conclusion or a claim that represents a person's belief about an issue or a case and evidence. The reason and objection elements are known as premises. The inference that represents the link to move from the evidence to the claim. A schematic representation of the simple argument structure is displayed in Fig. 1.

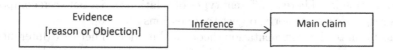

Fig. 1. A simple argument.

Different elements of the argument can organize to construct the argumentation model which provides the base of computer-supported argumentation tools. There are many argumentation models each of which can be applied to different contexts such as legal, philosophical, and political. Even though the models are different in their application, they share the basic argumentation principles such as establishing sound evidence of claim and reason, considering arguments and counter-arguments, and relating the facts to the proposed claim [28]. There are various argumentation models such as Wigmore model for solving the legal cases [30], Toulmin model widely used for teaching a critical thinking [23] as well as it used in psychological study and crisis' debate [34]. We will focus in the IBIS (Issue Based Information System) which was proposed by Horst Rittel and Webber for collaborative problem solving [30]. This model is simple, clear and easy to learn. The main objective of the model is to explore the different aspects of the Issue, convenes the others and supports constructive discussion [7]. It is used to make a

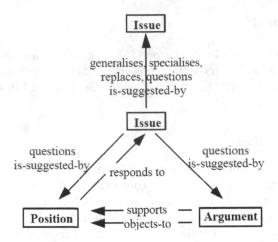

Fig. 2. IBIS model.

decision in political planning [8] and to support the community to make a decision [29]. In addition, it is used in architectural design and urban planning [7]. The ontologies of this framework are questions [issues], ideas [positions], argument pros and cons Fig. 2. Questions are the problems or the issues required to be brought up [6]. The position is the proposed solution or idea or the answer to the given question. It is always neutral. The arguments pro and con present opinions, facts, data and rhetoric. It uses as a way to evaluate the proposed ideas to give a solution for the question [8]. There is different types of relations exist between components such as responds to, supports, object-to, questions etc. [30].

In addition to the argumentation theoretical models. There are different argument visualization representations. Visualization plays a vital role in improving critical thinking, organizing thought, making reasoning and communication clearer, removing ambiguities, making argumentation more explicit for criticism, resolving disagreements resulting from of misunderstanding of the concept and helping to come up with a better decision [3]. There are many forms of the argumentation visualization such as linear, graph, threaded, container and matrix. Table 1 [28] presents representation style, typical uses and each type advantages and disadvantages.

2.2   Narrative Reasoning

Narrative or story-telling reasoning is a type of cognitive reasoning. Psychologists used it as an alternative to the regular argument approach for the evidence reasoning [4]. The aim of the narrative approach is to create cause-effect relation or a causal networks between events for guiding problem solving, and organizing human experiment [2,4]. Moreover, it is used for decision-making and problem solving, in which stories are useful for clarifying ambiguities. In addition, it plays a vital role for human senses making [36] because most of humanitarian-based

Table 1. Argument visualization (source: [28]).

Representation style	Typical uses	Pros	Cons
Linear (e.g., chat)	- Discussions (especially synchronous)	- Familiar and intuitive to most users, easiest to use	- Risk of sequential incoherence
		- Best to see temporal sequence and most recent contributions	- Not suited to represent the conceptual structure of arguments
		- Lack of overview	
Threaded (e.g., forums, academic talk)	- Discussions (especially asynchronous)	- Familiar and intuitive to most users, easy to use	- Moderately hard to see temporal sequence (because of multiple threads) as compared to Linear
	- Modeling	- Easy to manage large discussions	- Limited expressiveness (only tree-like structures)
		- Addresses issues of sequential incoherence	
Graph (e.g., Belvedere, Digalo)	- Discussions	- Intuitive for of knowledge modeling	- Hard to see temporal sequence
	- Modeling	- Highly expressive (e.g., explicit relations)	- Lack of overview in large argumentation maps; needs a lot of space; can lead to "spaghetti" images
		- Many graph-based modeling languages exist	
Container (e.g., SenseMaker, Room 5)	- Modeling	- Easy to see which argument components belong together and are related	-Limited expressiveness (e.g., only implicit relations, only tree-like structures)
			- Lack of overview in large argumentation maps because of missing relations
Matrix (e.g., Belvedere)	- Modeling	- Easy systematic investigation of relations	- Limited expressiveness (e.g., supports only two element types (row, column), no relations between relations)
		- Missing relations between elements are easily seen	- Uncommon (non-intuitive) way of making arguments

crises happen as a result of accumulative sequence of events and factors over time [21]. For instance, twistori[1] creates a real-time sequence of the tweets from twitter, BBC's user-Generated Content (UGC), which uses the social media for generating news, creates story to verify the different posted media such as photos, videos etc. [18].

2.3   Participant Geographic Information System (PPGIS)

A geographic information system (GIS) refers to the tools which focus on storing, displaying, analyzing and manipulating a geospatial data [32]. Nowadays, there

---

[1] http://twistori.com/.

Table 2. Participants based on different times and places.

	Same time	Different time
Same place	Community meeting: 2D, 3D, and animated project visualization, note keeping	Speaker series, shared Internet access: video recording, argumentation recording and structuring
Different place	Video conference, chat room: Shared text, graphics documents, virtual worlds	Internet newsgroups, forums, guest books: Argumentation recording and structuring, hyperlinking

is endeavoring to use volunteer data for updating county infrastructure and geographic changes caused by natural disaster. For example, public volunteered update Pakistan and Haiti map after the disaster [9]. The public power is not only concern with the updating map, but it extends to reports about emergency situations such as natural and human-made disasters. In which, local people are the first party responded to the problem [17]. As an example of this is during the Kenyans violence a public provided report about the incidents before media covers [13]. From different perspective, PPGIS concerns about involving public community in decision making process for the places they live in using the GIS [27,38]. An example of planning problem: water resource management, municipal, habitat site management, and solid waste management. Integrating public during the spatial planning process had many benefits. First, the time needed to resolve and find the best solution for the problem will be short. Second, local participant will be able to provide a better information about the social data and environment [27]. Third, integrate different individuals with different backgrounds in the planning process will cover many environments, economical and social factors. As a consequence, a reasonable solution will produce from the community [25] and the final decision can be more sustainable, effective and democratic [19]. Finally, involving the local citizens in the planning process assure the citizen to accept the consequence of the final decision [25].

Over the time, PPGIS is moved from using tradition methods such as public hearings, telephone enquiries, etc. to the Web-based for instance video chatting, voice conference etc. [5]. PPGIS discussion takes different forms based on the number of participants involve in the discussion, their space and/or time separation, the contribution types, and the existing of the moderator or his/her absent [11,24]. Table 2 [24] presents the different discussion methods base on the time and space. It is obvious from the table that Internet newsgroups, forums and guestbooks play vital roles when the decision process takes different time and has geographically disturbed participants.

Figure 3 [24] presents an argumentation mapping model proposed by Rinner [16,26] for mapping and integrating argumentation concepts with online geographic Information System (GIS) [31]. This conceptual model uses the spatial reference argument to facilitate and encourage participant discussion for decision

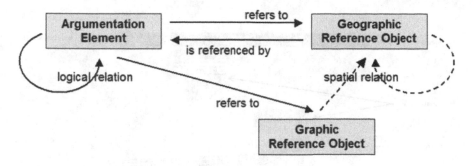

Fig. 3. Argumentation mapping.

making [1, 24, 27]. It consists of three main components which are argumentation element, geographic reference object and graphic reference object. The argumentation element is a self-reference relation for representing logical relation such as reply to a comment. Geographic reference object is also self-reference relation to represent the topological relation. Above all, this model indicates the relation between the geo-reference location and argument elements. Graphic reference object represents a new graphic location associated with argument elements added by deliberation's participant to explicit his/her idea.

PPGIS tools are varying from more structure to less structure, but all have the same objectives which is to involving public voices into the planning process. There are many PPGIS applications which can be categorized into two sub-categories. The first category integrate the threaded-based forum such as Argoomap, ArguMap, and GeoDF. The second category does not integrate the argumentation mapping for instance MAPPLER and Virtual Slaithwaite.

3   System Description

There are four platforms namely graph version, threaded-constrained version, threaded non-constrained version and narrative or story-telling version were implemented to address the proposed research questions. In this section we address each platform unique features and it is ontology individually.

3.1   Graph Version

This version uses the graph representation for evident reasoning and structuring the argument. The negotiation is presenting as a node-and-link predominantly for making the information explicit. The graph will be a directed graph from the reply node to the selected node. This version constrained the deliberation with the IBIS argument model. The relation between the nodes will be colored Fig. 4. The graph nodes include the discussion data such as video, image, text, etc. provided by the interacted users.

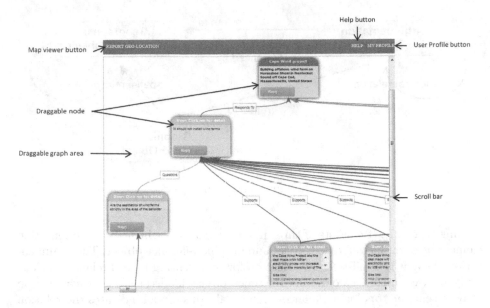

Fig. 4. Graph version interface.

3.2   Thread Versions

There are two versions of the threaded-visualization system that structured the deliberation as tree or a hierarchical order representation. The first system is constrained the discussion with IBIS model Fig. 5, while the second is a free-based discussion similar to online forum Fig. 6. For the constrained version, the messages types are present as a prefix to the message title. In addition, it will have different colors based on the IBIS scheme.

3.3   Narrative Version

Figure 7 presents the final version that is the narrative or story-telling environment. From the upper right there is a big drawable map which by default had a marker to the reported problem geo-location. Stories list tab has a list of all stories created for the reported incident. The stories' name is colored based on the opinion or the idea it reflects such as red refers to the stories against the idea of the reported incident, the green presents stories supports. Gray stories are just information about the issue. Finally the black colored are stories which did not classify by it is owner. When the user clicks on the story name, a sequence of events that the story consist of is presented in Story events' sequence area. The events which present as square colored box are numbered based on it order. They also can have an implicit or explicit geo-location associate with it. If the user click on an event which associate with geo-reference, the map will move and focus on the shape that represents the location.

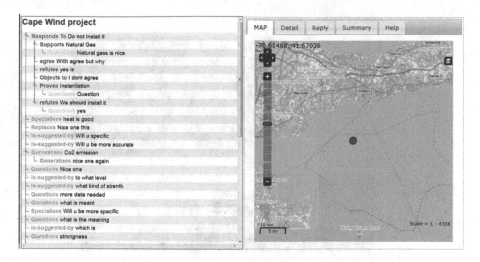

Fig. 5. Thread constrain interface.

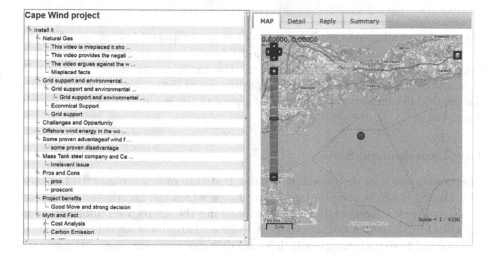

Fig. 6. Thread non-constrain interface.

4   Experiment and Results

The primary aim of the experiment is to investigate different environments for spatial problem knowledge accumulation. The four environments are three expository/argumentative discourse environments and one narrative environment. We believe that conducting an experiment is the best approach for addressing our proposed research questions. We test the effect of different argument or discussion representation for knowledge construction regarding wicked spatial problems. In addition, we are interested to investigate the effect of the constrained

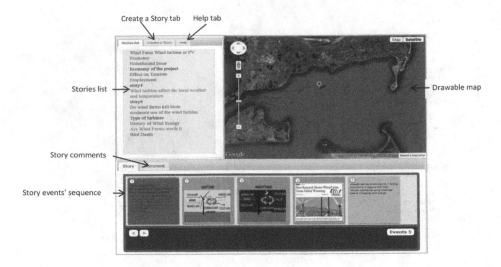

Fig. 7. Narrative/story-telling interface.

argumentative discourse environment on the participants' interaction. Finally, investigate spatial knowledge construction by considering two different cognitive reasoning modes. In Sect. 4.1, we present the experimental setup. Section 4.2, experiments results and our research questions are tested. Section 4.3, presents the results' discussion. Finally, Sect. 4.4 presents the evaluation of the different platforms.

4.1 Experimental Setup

The experiment was performed at Masdar Institute of Science and Technology, Abu Dhabi, UAE. It involved two main phases which are knowledge generation by using the developed platforms and evaluating the aggregated information in each platform based on two aspects knowledge and argument aspects. First phase consisted of 20 participants, in which 5 from different programs were assigned randomly to one of the proposed systems. The second phase of the experiment consisted of 17 participants who did not participate in the first phase of the experiment. In total, a community of 37 research engineers took part in this study.

Experiment Phase 1: At the beginning, twenty participants' took parts in this phase. Only sixteen volunteers posted messages in the online discussion, 13 were male and 3 were female. Ages ranged from 23 to 35 years old. The twenty participants had an individual session with the researcher about:

1. Argumentation, which basically focus in the IBIS model.
2. Brief introduction about the case study used in the experiment.
3. An instruction demo of using the system.
4. Participant registration in the system.

The participants were asked to discuss about "Cape Wind Offshore Energy Project" in Nantucket Sound, Massachusetts, USA. They were given reading material about the case. The deliberation topic was selected based on the following criteria: (1) It is a complex energy system, (2) It have geographic map location, (3) It provide various, heterogeneous and substantial amounts online-information sources such as Tweets, YouTube videos, images, blogs etc. and (4) It have to be controversial in which a stakeholder with different positions had various pro points and contra points.

The participants were assigned with a unique user name. In addition, their identity was anonymous. Participants were be able to access and view only the version they were assigned to. At the beginning, some participants had problems using the system. For that reason, some met the researchers for a second explanation of how to use the system. While, others created a second account because they forgot their passwords. During the experiments period which lasted for one week the participants received a daily email. The email identified the remaining period to finish the experiment and checked if the participants needed any helps. The participation's level was very low when the experiment started. Remarkably, the systems were online for one week and there was extension to some versions based on the participants' request. At the end of this phase the participants were asked to fill out a demographics questionnaire. The messages generated by the participants collaboratively in this phase provided the data which was used as the independent variables for addressing the research questions.

Experiment Phase 2: After finishing phase 1 of the collaborative work for producing the data in the four systems; the second phase start by asking a new group of participants to evaluate the generated data individually. Those users had an access to all of the four versions. In which, each participants had four identical evaluation forms to evaluate the data in each system. The aim of the evaluation form is twofold. First, to evaluate participants performance based on the knowledge produced; the evaluation criteria were based on the data collected indirectly from 47 press releases in a public hearings to the stakeholder about the main issues address for the cape wind project used in the experiment [20]. It consists of 14 categories: Economic analysis, Analysis of alternatives, Project goal and justification, Permitting process, Jurisdiction and authority, Energy source, Fuel diversity, Wind technology, Electricity rate change, Fiscal impacts, Environmental trade-offs, Socioeconomic impact, Health and safety impact, and Construction and removal operations. The participants' assessment of the scale two, four and six (2 = few, if any, identified, 4 = identified but unclear how they affect situation, 6 = clear and identify how they affect situation). Second, to assess the participants' according to Halpern's conception of arguments in which the evaluation's criteria were based on Halpern criteria for evaluating an argumentative quality [14]. It consists of the 7 categories: Quality of conclusions (claims), Premises are sound, Adequacy of premises, Assumptions related, Credibility of premises, Counterarguments accommodated, and Organization of arguments. The second evaluation criteria was applied for assessing the

expository or argumentative approaches which represent by graph, threaded constrain and threaded non-constrain platforms. The messages generated in the
phase 1 of the experiment consider the independent variables for the second
phases and users' score in the evaluation form are the dependent variables. The
individual scores were achieved by summing the number of points achieved in
each main category of the participant performance and the argument quality,
separately. The participants' scores provide the data for answering the research
question proposed in this work.

4.2   Results

We begin by presenting the number of messages posted in the expository environments. The users generate 27 posts in the graph versions, while the threaded
constrains users generate 70 posts while for the threaded non-constrains they
produce 42 posts. Apparently, the graph version users post the fewest number of
messages. This result is similar to that in the empirical study which concluded of
the effects of representational guidance on collaborative learning processes [33].
The threaded constrains environment produce the largest number of the messages. Regarding the narrative environment, there are 14 stories which consist
of 60 events in total.

Figure 8 presents the percentage of the content developed by the participants
of each category. Socioeconomic impact and environmental trade-offs were the
main issues addressed by most platforms.

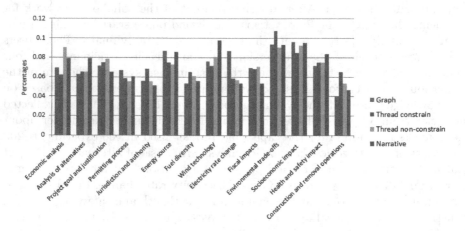

Fig. 8. Percentage of different categories generated by different versions.

Following is the descriptive analysis for participants' performance and argument's quality the mean and standard deviation of the participants' performance in the graph, threaded constrain, threaded non-constrain and narrative
are (52.71, 10.838), (36.12, 6.264), (59.65, 7.254), and (50.59, 10.880) respectively. From the result, we observe there is small diversion in the data generated

by the participants who used threaded constrain version, while there is a large variation in the graph and narrative platforms.

The expository environments, which are graph, threaded constrain and threaded non-constrain, were evaluated based on their argumentative quality aspects. An initial analysis of the ordinal variables (performance data and argument data) reveals that they are not normally distributed as shown by the histograms in Figs. 9 and 10.

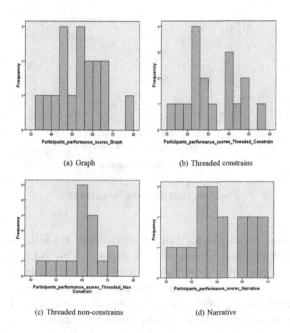

(a) Graph   (b) Threaded constrains

(c) Threaded non-constrains   (d) Narrative

Fig. 9. Participants' performance.

Now, we will study the proposed research questions. Our first research question dealt with differences in participant performance scores and argument quality scores of two types of argument environments: threaded constrain, which based on using the IBIS model, and threaded non-constrain. We conducted dependent-samples Wilcoxon tests on the participants' performance and argument quality scores between those two environments. There was significant effect of the constrained environment on the participants' performance, $z = -3.623$, $p < 0.05$. Results also showed significant effect of the constrained environment on the participants' argumentative quality, $z = -3.300$, $p < 0.05$. Our expectation that constrained environment affect participants' performance and argumentative quality was confirmed. The second research question was aimed at understanding differences in participants' performance and argumentative quality of discussion regards the discussion representation which is graph, and threaded constrains environment. Dependent-samples Wilcoxon tests of participants' performance and argument quality scores showed a significant difference between the

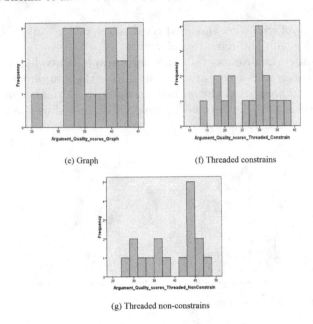

(e) Graph   (f) Threaded constrains

(g) Threaded non-constrains

Fig. 10. Participants' arguments.

graph and threaded constrain for both participants' performance, z $= -3.460$, p < 0.05, and participants' argumentative aspects, z $= -3.630$, p < 0.05. Our predication that argument representation affects participants' performance and argumentative quality was confirmed. The third question was aimed at exploring differences in participants' performance scores of discussion when participants used argument reasoning based on graph, threaded constrain and threaded non-constrain, and narrative reasoning. The Friedman test was conducted with scores as a dependent variable. There was a statistically significant difference in total score, $\chi^2(3) = 33.278$, p < 0.05. Post hoc tests indicated that there is a significant effect between the narrative version and the thread constrained version, z $= -3.627$, p < 0.05. Furthermore, there is a significant difference between the narrative version and threaded non-constrained version, z $= -2.854$, p < 0.05. Our expectation that reasoning modes affect participants' performance was confirmed.

4.3   Discussions

We begin by studying the effect of different reasoning and argument representation for harvesting the knowledge about the spatial planning problems. Socioeconomic impact and environmental trade-offs are the main issues addressed in the different environments as presented in Fig. 8; this may result from participants' background and their study in Masdar institute of Science and Technology.

Differences between the threaded constrain and threaded non-constrain in the way of participant performance and argument quality for exploring the topic of high controversial problem were found. The IBIS model in a threaded constrains version lead to reduce in the participant performance and argument quality scores. It may be easier for the participants into non-constrain threaded than participants in the threaded constrains to use the system because threaded constrain participants needed time to master the model in an efficient way as noted by one of the threaded constrain participant: "often needing to explore many things one by one but it was time-consuming".

The differences between the graph version and threaded constrain version in the both participants' performance and argument quality scores implies that the participants could construct better knowledge and argument when using the graph version. This could result from the positive aspects of using graph visualization in which it makes the information explicit. Therefore, it becomes

Table 3. Question 1.

Question 1: What did you like more and what did you not?	
Graph	1. I liked that we can add relationship to the previous comments, so we can have a rational idea about the whole topic and people opinions
	2. What I hated? Nothing in particular
	3. I liked the simplicity of the platform and how it was very well integrated to other internet search and social media services
Thread constrains	1. I liked the system as It was easy to use but some times the other members were not responding
	2. Like: The platform to express one's opinion in an anonymous way
	3. Not Like: The system is not very user-friendly
	4. It is a new concept for discussing things without revealing the users, very informative
	5. I did not like the no. of steps included for doing things
Thread non-constrains	1. Like: The topic was very interesting as it was related to the renewable energy
	2. Unlike: As we all are working in renewable energy sector therefore we already know the pros and cons of the wind energy. So it was hard to debate on the subject
	3. User interface navigation tool, few bug and design to fox
	4. There was not detail description about the project feasibility like cost aspect LCOE finance breakdown
	However a reasonable understandably can be taken from the available material
	5. Vibrant online discussion of views and counter-views in a civil manners
Narrative	1. I like the platform of discussion via stories and comments
	2. I did not like the input box and initial look of clarity for ideas/stories
	3. Way to create the story and flow of knowledge
	4. In terms of the features on the system, it is easy to use the drag drop features, search features provided that the user knows how to use it. I like the features but system is lack of giving such information in straight forward manner
	5. I like: Add story, Add comments
	6. I did not like: not to modify my story

Table 4. Question 2.

Question 2: What do you think is missing?	
Graph	1. I think we need to make the comments viewable in both graph and the typical list view
Thread constrains	1. I think the interface could be made more attractive
	2. Make the system more easy to use and give more platforms to express opinion
	3. The user manual/guide for using this discussion forum is missing
Thread non-constrains	1. More topics may be included
	2. There must be some true/false and MCQ to get the opinion of the users or to test their level of knowledge
	3. User access to add more discussion forum
	4. There was not detail description about the project feasibility like cost aspect LCOE finance breakdown
	However a reasonable understandably can be taken from the available material
	5. Active participation of every members
Narrative	1. Direction of use
	2. Creating new Topic
	3. Some error handling is missing such as once a story is created and mistakenly saved without a name, everything is gone, can't revert it back. It is necessary to provide some error saying, story name can't be empty

Table 5. Question 3.

Question 3: What do you think could be improved?	
Graph	1. It would be better to organize the view of the graph
	2. Speed of loading pages
Thread constrains	1. The interface can be improved
	2. Make this kind of system more widely used and enable the public to participate in the decision making process
	3. Providing user guides and more user-friendly look and feel GUI (Graphic User Interface)
Thread non-constrains	1. I think more topics, some detail summary, an interactive tutorial, some knowledge based questions
	2. Restriction is posting unrelated issue and comments and user should get update on the post under their post
	3. There was not detail description about the project feasibility like cost aspect LCOE finance breakdown However a reasonable understandably can be taken from the available material
	4. Few design aspect and integration with other social media like Facebook, twitter so that the no. of participants will increase
Narrative	1. Overall code and layout
	2. GUI, remove bug
	3. More straight forward design, and adequate information to know what to do, and error handling would make the system more useful, easy to use and user friendly
	4. Edit and deleting the story case

Table 6. Question 4.

Question 4: Overall experience in using the tools?	
Graph	1. I think the tools are very easy and straight forward to understand
	2. Very easy and convenient to navigate around
Thread constrains	1. Pretty good
	2. Overall opinion is this is nice way of communicating with public
	3. It was nice experiencing this tool, often needing to explore many things one by one s but it was time-consuming
Thread non-constrains	1. Good
	2. Very good experience. Hope get chance to participate in similar environment again
	3. Average
	4. Excellent
Narrative	1. Good
	2. Good experience, with fun and knowledge
	3. I had a good experience of using different tools in a single platform. I liked it
	4. Good

easily to analyze transcript, mediate and encourage debate, and make decision [28]. During knowledge construction for decision making, this representation explicit any gaps between the presented information. As a consequence, this help to address different aspects of under-take problem [15].

The four representations in our study (graph version, threaded constrain, threaded non-constrains and narrative) did result in significant differences in participants' performance and argument quality scores. The differences were between the narrative version and each of threaded constrains and non-constrains versions. While, there was no differences between the narrative version and the graph version. Apparently, this could result from the similarity between the narrative's representation and graph representation in the way of presenting the information.

In general, the results were subjectively affected by the number of the system users in phase one in which there were five participants takes part in it. In addition, the predominance of the competition between the participants in which they rewarded upon the number of individual participant posted messages in relative to others. This affected the quantity instead of the quality of the posted messages.

4.4 System Evaluation

For evaluating the proposed prototypes, we asked the participants to fill open ended questions about the system. The results are presented in the following tables: Tables 3, 4, 5 and 6. On average the participants were satisfied with the system.

5   Conclusions

In this work, we have investigated different knowledge construction frameworks for geo-spatial incidents. Four systems were implemented based on two cognitive thinking modes for sense-making, which are argument reasoning and narrative reasoning. Three of the proposed frameworks used the expository/argumentative discourse mode, while the fourth based on narrative reasoning mode. There are two main aspects were addressed when developing the argumentative discourse platforms. The aspects are argument presentation or visualization such as diagrammatic and threaded and the argument theoretical models. We proposed a series of research questions which could be addressed from the implemented frameworks. The main objective of this research is to test different environments to aggregate knowledge about geographically wicked problems. We have demonstrated that graph representation provides better results than threaded representation in the knowledge construction and in the quality of the argument. We also illustrate that the argument theoretical model leads to reduce participants' performance for producing data and argument quality scores. Moreover, we conclude that there is no significant difference between narrative representation and graph representation in the participants' performance to construct knowledge.

This works contributes in the PPGIS domain for knowledge construction in different ways. It is for the first time the PPGIS system integrates a diagrammatic representation for presenting the discussion and it integrates a new approach which is narrative reasoning for sense-making by integrating the story element which explicitly or implicitly associates with a geo-location. In addition, it integrates Web 2.0 mesh-up such as Facebook, YouTube, Blogs, News, etc. which not supported by any of the existing systems. There are two aspects for future improvement. The first aspect considers the system missing features and the second aspect considers the experiment limitations. Regards the first aspect, there are many features and extensions which could be made to improve the existing systems such as: study the influence when the users have the option to create a collaborative story, Create a reputation system such as voting and use text mining mechanism for automatic classification of the participant discussion. Regard the experimental limitations aspect, Instead of asking people to evaluate the content; we can use Ward's hierarchical clustering algorithm for clustering the discussion based on the contents, compute the Cohen's kappa for establishing inter-rater reliability of the coded messages. This value computed by a consensus agreement between two coders, increase the number of the participants for each version and choose a case study that relates and affect the system user.

References

1. Aaron, S., Claus, R.: A scalable GeoWeb tool for argumentation mapping. Geomatica **65**(2), 145–156 (2011)
2. Artz, J.M.: Narrative vs. logical reasoning in computer ethics. ACM SIGCAS Comput. Soc. **28**(4), 3–5 (1998)

3. Austhink: argument mapping tutorials (2006). http://austhink.com/reason/tutorials/
4. Bex, F., Prakken, H., Verhey, B.: Anchored narratives in reasoning about evidence. In: Proceedings of the 19th Annual Conference on Legal Knowledge and Information Systems (JURIX), pp. 11–20 (2006)
5. Butt, M.A., Li, S.: Developing a web-based, collaborative PPGIS prototype to support public participation. Appl. Geomatics **4**(3), 197–215 (2012)
6. Conklin, J.: Dialogue Mapping: Building Shared Understanding of Wicked Problems. John Wiley and Sons Inc., New York (2005)
7. Conklin, J., Begeman, M.L.: gIBIS: a hypertext tool for exploratory policy discussion. In: Proceedings of the 1988 ACM Conference on Computer-supported Cooperative Work (CSCW), pp. 140–152 (1988)
8. Conklin, J., Begeman, M.L.: gIBIS: a tool for all reasons. J. Am. Soc. Inf. Sci. **40**(3), 200–213 (1989)
9. Crooks, A.T., Wise, S.: Modelling the humanitarian relief through crowdsourcing, volunteered geographical information and agent-based modelling: a test case - Haiti. In: Proceedings of the 11th International Conference on GeoComputation, pp. 183–187 (2011)
10. Dangermond, J.: Living Maps — Making Collective Geographic Information a Reality. O'Reily, Santa Clara (2011)
11. Floreddu, P., Cabiddu, F., Pettinao, D.: Public participation in environmental decision-making: the case of PPGIS. In: D'Atri, A., Ferrara, M., George, J.F., Spagnoletti, P. (eds.) Information Technology and Innovation Trends in Organizations, pp. 37–44. Physica-Verlag HD, Heidelberg (2011)
12. Gürkan, A., Iandoli, L., Klein, M., Zollo, G.: Mediating debate through on-line large-scale argumentation: evidence from the field. Inf. Sci. **180**(19), 3686–3702 (2010)
13. Heinzelman, J., Waters, C.: Crowdsourcing crisis information in disaster affected Haiti. Technical report, United States Institute of Peace (2010)
14. Jonassen, D.H.: Learning to Solve Problems: An Instructional Design Guide. Pfeiffer, San Francisco (2003)
15. Kanselaar, G., Erkens, G., Andriessen, J., Prangsma, M., Veerman, A., Jaspers, J.: Designing argumentation tools for collaborative learning. In: Kirschner, P.A., Buckingham Shum, S.J., Carr, C.S. (eds.) Visualizing Argumentation, pp. 51–73. Springer, London (2003)
16. Keßler, C., Rinner, C., Raubal, M.: An argumentation map prototype to support decision-making in spatial planning. In: Proceedings of the 8th AGILE Conference on Geographic Information Science (AGILE), pp. 135–142 (2005)
17. Malizia, A., Bellucci, A., Diaz, P., Aedo, I., Levialdi, S.: eStorys: a visual storyboard system supporting back-channel communication for emergencies. J. Vis. Lang. Comput. **22**(2), 150–169 (2011)
18. Meier, P.: Verifying crowdsourced social media reports for live crisis mapping: an introduction to information forensics (2011). http://irevolution.files.wordpress.com/2011/11/meier-verifying-crowdsourced-data-case-studies.pdf
19. Meng, Y., Malczewski, J.: Web-PPGIS usability and public engagement: a case study in Canmore, Alberta, Canada. J. Urban Reg. Inf. Syst. Assoc. **22**(1), 55–64 (2010)
20. Mostashari, A.: Collaborative Modeling and Decision-Making for Complex Energy Systems. World Scientific Publishing Company, New York (2011)
21. Okolloh, O.: Ushahidi, or 'testimony': web 2.0 tools for crowdsourcing crisis information. Participatory Learn. Action **59**, 65–70 (2009)

22. Rahwan, I., Zablith, F., Reed, C.: Laying the foundations for a world wide argument web. Artif. Intell. **171**(10–15), 897–921 (2007)
23. Reed, C., Rowe, G.: A pluralist approach to argument diagramming. Law, Probab. Risk **6**(1–4), 59–85 (2007)
24. Rinner, C.: Computer support for discussions in spatial planning. In: Campagna, M. (ed.) GIS for Sustainable Development, pp. 167–180. CRC Press, Boca Raton (2005)
25. Rinner, C., Bird, M.: Evaluating community engagement through argumentation maps—a public participation GIS case study. Environ. Plan. B: Plan. Des. **36**(4), 588–601 (2009)
26. Rinner, C., Keßler, C., Andrulis, S.: The use of Web 2.0 concepts to support deliberation in spatial decision-making. Comput. Environ. Urban Syst. **32**(5), 386–395 (2008)
27. Rinner, C., Kumari, J., Mavedati, S.: A geospatial web application to map observations and opinions in environmental planning. In: Advances in Web-based GIS, Mapping Services and Applications, pp. 277–291. CRC Press (2011)
28. Scheuer, O., Loll, F., Pinkwart, N., McLaren, B.M.: Computer-supported argumentation: a review of the state of the art. Int. J. Comput.-Support. Collaborative Learn. **5**(1), 43–102 (2010)
29. Schneider, J., Groza, T., Passant, A.: A review of argumentation for the social semantic web. Seman. Web **4**(2), 159–218 (2013)
30. Shum, S.B.: The roots of computer supported argument visualization. In: Kirschner, P.A., Shum, S.J.B., Carr, C.S. (eds.) Visualizing Argumentation, pp. 3–24. Springer, London (2003)
31. Sidlar, C.L., Rinner, C.: Analyzing the usability of an argumentation map as a participatory spatial decision support tool. J. Urban Reg. Inf. Syst. Assoc. **19**(1), 47–55 (2007)
32. Simao, A., Densham, P.J., Haklay, M.: Web-based GIS for collaborative planning and public participation: an application to the strategic planning of wind farm sites. J. Environ. Manag. **90**(6), 2027–2040 (2009)
33. Suthers, D.D., Hundhausen, C.D.: An experimental study of the effects of representational guidance on collaborative learning. J. Learn. Sci. **12**(2), 183–219 (2003)
34. Voss, J.F.: Toulmin's model and the solving of ill-structured problems. Argumentation **19**(3), 321–329 (2005)
35. Weil, A.A., Ivers, L.C., Harris, J.B.: Cholera: lessons from Haiti and beyond. Curr. Infect. Dis. Rep. **14**(1), 1–8 (2012)
36. Yearwood, J., Stranieri, A.: Narrative-based interactive learning environments from modelling reasoning. J. Educ. Technol. Soc. **10**(3), 192–208 (2007)
37. Zarefsky, D.: Argumentation: The Study of Effective Reasoning, 2nd edn. The Teaching Company, Chicago (2005)
38. Zhao, J., Coleman, D.J.: GeoDF: towards a SDI-based PPGIS application for E-governance. In: Proceedings of the GSDI-9 Conference, pp. 1–14 (2006)

Author Index

Printed in the United States
By Bookmasters